The Dating Game

How old is the Earth? At the end of the nineteenth century, geologists, biologists, physicists and astronomers were all looking for a clock that would provide an answer to this, the greatest Time question of all.

The Dating Game tells the story of one man's vision of developing a geological timescale that would finally lead to an accurate date for the age of the Earth. Despite scientific opposition, financial hardship and personal tragedy, Arthur Holmes, the greatest geologist of the twentieth century, fought for fifty years to convince The Establishment of an Earth of great antiquity: a fight which eventually transformed the moribund 'art' of geology into a dynamic science.

Cherry Lewis' engaging writing brings Holmes back to life and skilfully weaves his adventures, loves and losses around the early history and science of dating the Earth, and the discovery of radioactivity – the clock that tells geological time.

Cherry Lewis has had a life-long passion for rocks and fossils. After Drama College and running her own retail business for several years, she finally decided to indulge her passion and studied geology as a mature student at Bristol University. Her degree was succeeded by a PhD on Tibetan granites and post-doctoral research at University College, London, which led to a career in the oil industry. Whilst writing her thesis she discovered the work of Arthur Holmes and became captivated by the man who pioneered the way towards determining the true age of the Earth.

The Dating Game

One Man's Search for the Age of the Earth

CHERRY LEWIS

It is perhaps a little indelicate to ask of our Mother Earth her age,
but Science acknowledges no shame.

CAMBRIDGE
UNIVERSITY PRESS

PUBLISHED BY THE PRESS SYNDICATE OF THE UNIVERSITY OF CAMBRIDGE
The Pitt Building, Trumpington Street, Cambridge, United Kingdom

CAMBRIDGE UNIVERSITY PRESS
The Edinburgh Building, Cambridge CB2 2RU, UK
40 West 20th Street, New York, NY 10011–4211, USA
10 Stamford Road, Oakleigh, VIC 3166, Australia
Ruiz de Alarcón 13, 28014 Madrid, Spain
Dock House, The Waterfront, Cape Town 8001, South Africa

http://www.cambridge.org

First published 2000

Printed in the United Kingdom at the University Press, Cambridge

Typeface Quadraat System QuarkXPress®[w21]

A catalogue record for this book is available from the British Library

Library of Congress Cataloguing in Publication data

Lewis, Cherry, 1947–
    The dating game : searching for the age of the Earth / by Cherry Lewis.
        p. cm.
    "It is perhaps a little indelicate to ask of our Mother Earth her age, but science
    acknowledges no shame."
        Includes bibliographical references.
        ISBN 0 521 79051 4
        1. Earth – Age.   2. Geological time.   3. Holmes, Arthur, 1890–1965.   I. Title.
    QE508 .L48 2000
    551.7'01–dc21   00–023615

ISBN 0 521 79051 4 hardback

To Steve
without whom this book
would not have been written

Contents

List of Illustrations

Prelude to The Game

To the reader who wishes to see something of the
'wild miracle' of the world we live in through the eyes of
those who have tried to resolve its ancient mysteries.

Arthur Holmes

I have always been a collector. I blame my parents. As very small children my sisters and I would be taken on warm sunny after-noons to the beach we overlooked from our house in Devon. Hours would be spent sifting through the sand and debris hunting for shells, the prize of which was the cowry. Rare and elusive, not much bigger than my thumb nail, the exotic, white-lipped and pink-backed shell, sometimes dotted with brown spots and sometimes not, was the greatest treasure on the beach. Somewhere in the attic a boxful of cowries awaits my retirement when I shall use them to make shell pictures and decorate little wooden boxes. These I shall give to my grand-children for Christmas who will give me a kiss then hide them in a cupboard with embarrassment: 'Oh, it was just something Granny made.'

From then on I was addicted. I walked around with my eyes on the ground just in case I missed some treasure – an unusual stone, a rare wild flower, a pretty feather, a sixpenny bit; nothing was overlooked and I collected it all. Aged eight and we were living in Iraq. We went for walks on the edge of the desert and one day I picked up a crimson stone. On the polished face a 'C' was clearly etched in orange. C for Cherry – my stone. 'God must have put it there for you to find,' my mother said. But I wasn't sure about God, so I held the stone in one hand and stretched out both arms before me. 'If there is a God', I told him telepathically, 'knock the hand holding the stone'. I waited and waited but nothing happened, he didn't knock my hand or smite me down for my blasphemy, but when I got home I couldn't find the stone. I had held it in my hand all the way back, and now suddenly it wasn't there anymore. I can feel the bitter disappointment even now. Where had it gone? Had God spirited it away to teach me a lesson? It seemed likely, it would be just like him. But that stone remained in my heart.

With my nose to the ground it didn't take long before I discovered fossils. My favourites were ammonites, the ribbed, tight, spiral whirls that came in such incredible shapes and sizes. Some would be coated in glistening cubes of fool's gold; some cracked open to reveal secret chambers filled with shimmering crystals; others as solid as stone had beautiful, fern-like patterns etched all over them. A few were enormous – three feet across. Occasionally a complete one just lay there waiting to be picked up, but usually I spent hours chipping and scratching away at the mud in the hope of revealing a jewel.

My geography teacher liked fossils too. He would take us out on field trips and explain that ammonites were the remains of squid-like creatures that had floated about in the warm shallow sea that covered southern England in Jurassic times, about 180 million years ago. Dinosaurs had roamed there too, but somehow they didn't interest me much, those huge lumbering giants that have excited so many. Much more enticing were discoveries at low tide of flattened, almost paper-thin ammonites, wet and iridescent, all colours of the rainbow. The mud they were buried in was so fine that it had preserved the original mother-of-pearl-like material their shells were made of, instead of turning it to stone. So those warm tropical seas had been full of colour even then, just like they are today. I put an ammonite on a chain and wore it around my neck, impressing my friends with its '180 million years'. But how did I know how old it was? Well my teacher told me of course, and so it just was.

Years later I became a geologist. Standing in a white, sterile laboratory I held in my hands cold, hard rocks that would reveal to me the secrets of the Himalayas – how long had they been there, why were they so high, how fast were they growing? Locked up within the tiny crystals that made up the rock were the answers I sought, I only had to reach in and prize them free. Measuring time in millions and even billions of years became routine, much like a banker gets used to dealing in millions of pounds, and an astronomer in millions of light years. By then I understood how to date a rock and why it is important to know, but not once had I ever thought to ask 'who?'. Who had shown the way? Who had dated the very first rock, and how had they done it?

Robert Shackleton, Professor of Geology and nephew of the great explorer Ernest Shackleton, was retiring. At 76 he had led our expedition to Tibet, striding up and down mountains at 15 000 feet when I could barely walk at that altitude, but now it was time he took things a bit easier. As he cleared his office he placed a large pile of unwanted papers on my desk. It was the last thing I needed, out-of-date papers, but I thanked him anyway. Distractedly I glanced through them and noted half a dozen signed by the author, a certain Arthur Holmes of whom, I admit, I had not heard. Perhaps it was the title that made me open up *The Age of the Earth*; perhaps it was the date '1913' that intrigued me. Whatever it was I started to read and was instantly captivated. It was then I found out 'who'.

Time is crucial to us all; in this frenetic world we live in we have a constant need to refer our lives to a time scale we understand, and for most of us today that scale is the twenty-four hour clock we wear on our wrists. At the end of the nineteenth century, however, geologists, biologists, physicists and astronomers were looking for another clock that would provide an answer to one of the greatest Time problems of all: How old is the Earth? Ingenious methods for measuring it were proposed but few came close to the truth because no accurate scale had been developed to quantify geological time.

At that time, understanding geology was like understanding history, but without the dates. Imagine history without any dates: you know that the Romans invaded Britain prior to the First World War, but you have no idea when – was it two million, two thousand or only two years before? You know that Kennedy was shot after the Second World War, but again, you don't know how long after. And when were those wars anyway? Without a scale of time there would be no way of knowing. With no dates you can only understand the order in which historical events happened, but not *when* they happened. Such was geology at the beginning of the twentieth century: history without any dates. How old were the first fossils, the first birds, first trees, and all those enormous dinosaurs? No one had any idea. All they knew was that the ones at the top were younger than the ones at the bottom, because that was the order in which the rocks were laid down: slowly and gradually one of top of another, so the oldest had to be at the bottom. Other than that there was no scale with which to measure geological time.

So why is it important to know the age of the Earth? What difference does it make to our lives? Stephen Hawking explains it well:

ever since the dawn of civilisation, people have not been content to see events as unconnected and inexplicable. They have craved an understanding of the underlying order in the world . . . our goal is nothing less than a complete description of the universe we live in.

By knowing the age of Earth rocks, Moon rocks and rocks from other planets we contribute to that 'complete description' and are more able to understand our place in the order of things, our relationship to other celestial bodies. It helps us to navigate our way around the Universe and to build up a picture of why we are here at all.

But it is not just finding our place in the heavens when dating rocks comes into its own. In the mid-1960s our ability to determine very precisely the age of certain rocks found on the ocean floor became a 'Eureka' moment in the history of science which, within a couple of years, led to the confirmation of the 'Theory of Plate Tectonics'. This theory now ranks as one of the great unifying theories of all time, alongside Darwin's theory on the Origin of Species and Einstein's General Theory of Relativity, explaining as it does just about every natural process that we ever observe on this Earth. But, as we shall see, without the ability to date rocks, we might still be in a position of speculation regarding this theory.

Today we state with authority that the Earth is four and a half billion years old, and some say with great confidence that '65 million years ago' an enormous meteorite collided with the Earth wiping out those dinosaurs and most of life. But how do we know those dates? How do we count all those years?

If you ask most people how geologists date the age of the Earth they will think for a bit, remember the Turin Shroud, and invariably answer 'carbon dating'. Not so. Carbon is the archaeologist's clock and can only be used to date things that are less than fifty thousand years old, which is of little use to the geologist. No, geologists have their own clock. It is made of uranium and lead, and has been ticking away ever since the Earth was formed in a nebulous cloud of dust. In order to find out how long ago that was, all we had to do was learn how to tell the time using this special clock. But first we had to find the clock.

For nearly 50 years the English scientist, Arthur Holmes (1890–1965) pursued these goals almost single-handedly. Indeed, so intertwined is the life of Arthur Holmes with his search for a geological clock and the age of the Earth that it is impossible to separate the two, thus the story of one becomes a biography of the other.

So, now let me introduce you to the players in the Dating Game, but before I do I should perhaps warn you: wild miracles occur throughout this book. Some are easy, some are hard, but they all require a little thought. Be brave.

A Brief History of Time

The age of the Earth has been one of the most controversial
numbers in science since the 17th century.

Stephen Brush

Primrose Hill in Gateshead was a modest street of single-bay
Victorian brick houses, terraced in tiers down the steep hill of
Low Fell. If you stood in the middle of the road the view below
was of green fields and a large sky, despite the town's location
in the industrial heartland of northern England, but the houses
were sideways on to this view and austerely faced each other
across the road, their front doors guarded from the street by
three feet, the occasional hydrangea, and an iron railing. In
January 1900, in the wintry dawn of a new century, Arthur
Holmes was ten years old and living at number nineteen, the
only child of staunchly Methodist parents. His father was a cab-
inet maker and worked as an assistant in an ironmonger's shop.
Consequently they were of modest means.

Not far away was Gateshead Higher Grade School where
Mr John Bidgood, the school's visionary headmaster, teacher of
biology, and world expert on tropical orchids, made sure that
provisions for the teaching of science were not exceeded by any
other municipal school in the country. It was, for example, the
first of those schools to have science laboratories specifically
designed and fitted for that purpose. In 1901, the year that Queen

19 Primrose Hill, Gateshead, where Arthur Holmes lived as a boy.

Victoria died, shy and retiring young Arthur Holmes joined this exhilarating school environment, and blossomed. By the age of fifteen he had won first class honours in the School Certificate Examinations, and so could go on to study in the sixth form. Walking to school every day with his nose in a book, he must have exemplified the school motto taken from an ancient Greek poem: 'Toil is no disgrace, But idleness is a thing of reproach'

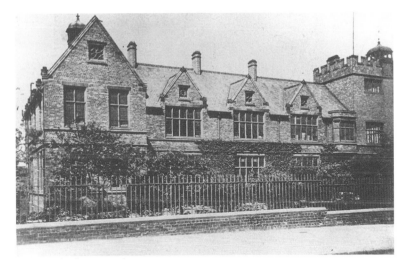

Gateshead Higher Grade School attended by Arthur Holmes
and Bob Lawson.

which was neatly, but perhaps unfortunately, rendered down to 'Toil no soil'.

Tragically Mr Bidgood died aged fifty-one, the year that Arthur passed his School Certificate, but he was replaced by Mr Walton, under whose headship musical education was greatly advanced. This too benefited Arthur, for as well as being academically outstanding he was an exceptionally talented pianist and frequently accompanied the school's Operatic Society in their famed performances of Gilbert and Sullivan.

Arthur's best friend was Bob Lawson, who lived at 19 Abbey Street, also in Gateshead. Unusually for those times, both boys were only children so they became to each other the brothers they had never had, inseparable companions both at school and at home. Arthur was just three months older than Bob so they entered the sixth form together and it was here that they fell under the spell of an inspirational physics teacher, Mr James McIntosh. What was so exceptional about Mr McIntosh was that not only did he manage to instil in the boys a desire to study

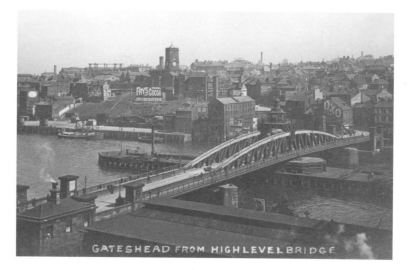

General view of Gateshead around 1910.

physics, in itself a quite remarkable achievement, but he taught them physics through other subjects so that they hardly realised they were learning it. They learnt physics in geology through earthquake activity, mountain building and volcanic eruptions; astronomy taught them about gases and solids, velocity and motion, and even history played a part as he told them marvellous stories about the discoveries and controversies involving famous physicists. But there is nothing more guaranteed to make learning exciting than a feeling of being 'involved', and it so happened that one of the most dramatic times in scientific history was unfolding right there and then.

Lord Kelvin had been the foremost physicist of his day and for many years the leading protagonist in the 'Age of the Earth' debate, an extraordinary scientific controversy that had endured for the past fifty years. But as Arthur and Bob studied for their Higher Certificate Kelvin was an old man of eighty-two and his theory, that the Earth was only twenty million years old, was being torn asunder by the young physicists of the day who were

An ironmonger's shop in Gateshead around 1910, similar to the
one in which Arthur Holmes' father worked.

working on the newly discovered phenomenon of radioactivity.
Kelvin found it difficult to accept that this work, into which he
had put so much effort, was being disposed of so readily. He
clung to his old ideas, getting more and more exasperated, until
one day in August 1906 while he was on holiday in France, he
read in The Times a report of experiments with radioactivity,
performed by Sir William Ramsay and Frederick Soddy, that
finally seemed to disprove his ideas.

Kelvin wrote an angry letter to The Times criticising their work.
While he recognised how 'brilliantly interesting' and 'solidly
instructive' it was, he also considered it added nothing new to
the theory of atoms that had been proposed by Democritus two
and a half thousand years ago! The irritation felt by the younger
generation at Kelvin's obstinate refusal to accept the new evidence was thinly veiled: 'Lord Kelvin's letter will of course

receive respectful attention . . . but it is also known that his brilliantly original mind has not always submitted patiently to the task of assimilating the work of others by the process of reading'. The correspondence continued in this vein for over a month. Arthur and Bob, on their summer holidays, were on the edge of their seats with the excitement of it all, for not only did they become familiar with all the arguments, they also got to know all the big names in science at that time – William Ramsay, Ernest Rutherford, Frederick Soddy and Robert Strutt (whom Kelvin pointedly referred to as 'Mr.' Strutt to make sure that no-one mistook him for his more illustrious father 'Lord' Strutt). Once again they were learning physics without realising it, and enjoying every minute of it. To read of these famous people discussing their science in this manner through the pages of The Times was a revelation to the boys; they felt they were watching history unfold before their very eyes; they felt 'involved'. No wonder they both decided to study physics at University.

Lord Kelvin was by no means the first person to try and date the age of the Earth. With insatiable curiosity Man had been trying for centuries to discover her carefully guarded secret. In 1650 it was widely accepted that, as stated in the Bible, God had created the World (and the heavens, then considered part of the World) in six days. So James Ussher, the Archbishop of Armagh in Ireland and a well known biblical scholar, decided to try and calculate not only the year in which God (the Christian God) had created the World, but also in which month, and on what day of the week. He worked it out by adding up the ages of all the important people mentioned in the Bible who had lived since the time of Adam, the first person created by God, and established that the World was created four thousand and four years before the birth of Christ, on the evening of the

Lord Kelvin in 1899, aged 75.

22nd October, which was a Saturday. This date, 4004 BC, for the Creation of the World as determined by Ussher was then printed as a note in Genesis, the first chapter of the Bible, whereupon it became 'gospel' and accepted without question in Christian teaching for several centuries.

Other religions had very different ideas. Zoroaster, a Persian prophet who lived in the sixth century BC, believed that the world had been in existence for over twelve thousand years; the Roman writer Cicero relates that the venerable priesthood of Chaldea in ancient Babylonia held the belief that the Earth emerged from chaos two million years ago, while the old Brahmians of India regarded Time and the Earth as eternal. It seems that Ussher's famous estimate probably represents the shortest period ever assigned to the age of the Earth. However, at that time Christianity in Europe was a powerful force, and a literal inter- pretation of the Bible dominated the understanding of science and the way people looked at the natural world. The result was a distorted explanation of geological phenomena as naturalists tried to cram millions and millions of years of Earth history into less than 6000 years, and those few individuals whose observations suggested that this interpretation might not be scientifically reliable were branded as heretics.

Another big geological problem for biblical scholars was the explanation of fossils. For thousands of years, ever since the Greeks and probably long before, people had observed objects trapped in rocks and wondered what they could be. For exam- ple, Leonardo Da Vinci, writing five hundred years ago, accu- rately deduced how fossils are preserved by closely observing what happened around him: he saw shapes in the rocks that looked like shells found on the sea shore, he saw rivers carry- ing large volumes of mud down to the sea, and by putting two and two together he deduced that fossils were once organisms living in the sea that had been buried by mud from rivers and turned to stone. He made sense of the world around him with- out needing to invoke magic vapours or the works of the devil, which was how many people in the Middle Ages explained fossils. But even when it was widely recognised that fossils were the remains of creatures that had once lived on the Earth or in the sea, it was nevertheless still problematic to find the remains of fishes and other sea-dwelling creatures on the tops of mountains.

Once again the answer came from a literal interpretation of the Bible. The story of Noah's Flood tells how God saw that the wickedness of Man was great, that He repented having made him in His image and so decided to destroy him. Only Noah was good enough to be saved. So God told Noah to build an Ark in which he was to take his wife, his sons and his sons' wives, plus a pair of every living thing on the Earth because He was going to cause it to rain for 40 days and nights and, in a fit of pique, destroy every living thing. So, even though Noah was 600 years old at the time, he built the Ark as God told him to and saved 'Life' for future generations.

The waters rose up and covered the mountains and every living thing was destroyed. The violence of the Flood was so great that the bottom of the oceans was stirred up, all soil was washed off the land, they were mixed together and then redeposited, even on the top of mountains, to form the stratified rocks we see today containing the remains of all the poor creatures who perished. And thus we got fossils on the top of mountains. With this explanation all rocks and fossils were supposedly laid down at the same time, and adherents to the 'Noachian Deluge' hypothesis dominated geological thought until the end of the eighteenth century.

There were, however, a few who attempted to explain the Flood by means other than the wrath of God. In 1696 William Whiston wrote a 'New Theory of the Earth' in which he explained how, on the 2nd of December, 2926 BC, a comet a quarter of the size of the Earth cut the plane of the Earth's orbit only nine thousand miles away. Whiston considered that the effect of this comet passing so close to the Earth generated a giant tide in the sea and in waters within the Earth; thus were the 'fountains of the great deep broken up' and unleashed upon the Earth. As the comet passed, water was discharged from its tail, which was when it rained for 40 days. According to Whiston, the result of all this water on the Earth was a flood six and a quarter miles deep. Undoubtedly a comet passing that close would wreak havoc on the Earth's weather and tides and Whiston's idea seems as plausible as that of a meteorite impact wiping out the dinosaurs.

Whiston however, could not prevail over the biblicists, so the Noachian Deluge theory triumphed and did even more harm to the development of geology as a science than Archbishop Ussher. Gradually though, a few men, understanding the significance of what they observed in the rocks, were prepared to stand up to the theologians, and an adherence to a strict interpretation of the Bible slowly gave way to the need for longer and longer periods of time to explain geological and biological processes.

In 1785 James Hutton, frequently called the 'father of modern geology', stood before the learned and recently formed Royal Society of Edinburgh. He read to them an essay he had written entitled 'Theory of the Earth' in which he emphasised the immensity of geological time and the uniformity of geological processes which, over vast time periods, formed the Earth as we see it today. Hutton was scoffed at by his critics for 'running about the hill-sides with a hammer to find out how the world was made', but it was because of this long and critical study of rocks in the field that he recognised that the Earth we are standing on now must have been made from the rocks of past ages. He explained to his audience how the land of today had been fashioned by the seas and rivers of yesterday, and that the land of tomorrow was today forming at the bottom of the sea. He overthrew much of current thinking, which he knew would arouse theological opposition, and identified at least three cycles of land formation in the Earth's history, each of which were 'of indefinite duration'. He concluded his lecture with the famous words: 'The result therefore, of our present enquiry is that we find no vestige of a beginning, – no prospect of an end'. He considered that 'with respect to human observation' geological time was simply too long for us to imagine. Being so flatly in contradiction with the Scriptures, this led to him being accused of having 'deposed the Almighty Creator of the Universe from his Office', but Hutton was insistent: 'In nature' he writes, 'we find no deficiency in respect of time'.

James Hutton died in 1797 before his theory had gained much credence, the year that Charles Lyell was born, a man destined to become one of the greatest influences in modern geology. When Lyell became a geologist (he originally trained to be a lawyer) he took up where Hutton had left off. He looked at the geological processes operating around him and realised that they were the key to understanding what had happened in the past. He considered that nothing could have occurred in the geological record that was not happening now, and that all that was required was a vast amount of time for geological processes to endlessly recycle and reshape the planet, eventually creating the world as we observe it today. But he had one significant advantage over Hutton.

In the 1790s William Smith, then a surveyor working on building canals to carry coal, recognised that there was a regular and systematic order within the rocks of southern England. A self-educated man, Smith observed that not only were rocks ordered in a way that could be followed for miles across the country, but that the suites of fossils contained within those rocks 'always succeed one another in the same order' making it possible to correlate one rock with another that contained the same suite of fossils, even though they were miles, or even countries, apart. But Smith was unable to reconcile what he observed with an understanding of how it had occurred, so, resigning himself to this state of affairs, he got on with his work, believing that these matters were not of his concern: 'I have left off puzzling about the origin of Strata and content myself knowing that it is so . . . The whys and wherefores cannot come within the Province of a Mineral Surveyor.' Nevertheless, he certainly had a feel for the immensity of time that was reflected in the rock strata he surveyed, declaring: 'The time required for [each] Perfection and Decay and subsequent formation into Strata . . . would stagger the faith of many'.

Smith made the first geological map of the 'Strata of England and Wales' which, had he been able to publish it immediately upon completion, would have put him closer to the forefront of geological fame. But he did not publish until 1815, and during the time that elapsed between making it and publishing it he discussed his ideas widely with his contemporaries such that the information he possessed became diffused throughout the geological community by word of mouth, contributing significantly to the progress of that science, but without its author getting due recognition. So by the time Charles Lyell was a geologist this crucial understanding that fossils allowed the rocks to be ordered one above the other in a chronological sequence had become an accepted meme.

During the course of his work Lyell examined the great volcano of Etna on Sicily and studied the historical records of its frequent eruptions. He noticed how each time it erupted a new layer of lava would be added on top of the previous one, causing the mountain to grow at a measurable rate. So by knowing the total height of the volcano, its approximate rate of growth and the frequency of eruptions, Lyell realised that it should be possible to estimate the age of the volcano. He did the calculations and determined that it must be several hundred *thousand* years old. While this in itself was an astounding revelation, the question remained, how much of geological time did a hundred thousand years represent? Had the volcano been growing ever since the world began, or was it a recent phenomenon? At the edge of the volcano, underneath the first lava flows, Lyell found fossil shells that were virtually identical to the shells of molluscs swimming in the Mediterranean at that time. From this he deduced that the fossils were geologically 'recent', that a hundred thousand years was geologically short, and that the age of the Earth must be immense.

Lyell's approach to geology was persuasive and he soon became an important influence over many scientific figures of the time, one of whom was Charles Darwin. It is not always appreciated that Darwin was first and foremost a geologist. Having been fascinated by the subject as a boy he went from finding formal instruction on geology 'intolerably dull' to being a leading member of the Geological Society on his return from the voyage of the Beagle. Four out of every five pages of notes taken on the Beagle's voyage were on geological topics, and on the few occasions that Darwin referred to himself as a scientist, he called himself a geologist.

From the time Darwin read Lyell's newly published book on board the Beagle in 1831, he recognised the significance of Lyell's work: 'I had brought with me the first volume of Lyell's 'Principles of Geology', which I studied attentively: and this book was of the highest service to me in many ways. The very first place which I examined showed me clearly the wonderful superiority of Lyell's manner of treating geology'. Within days of his return from the voyage Darwin got in touch with Lyell and they became lifelong friends. Through his geological observations, Lyell provided Darwin with the unfathomable amounts of time required to unfold the evolution of life. In the process Lyell influenced not only Darwin's geological conclusions, but ultimately those on the origin of species, because, in the words of Thomas Huxley, Darwin's great champion, 'biology takes its time from geology'.

So at this point geology seemed to have turned full circle. From the theological time restrictions of 6000 years for the age of the Earth imposed by Archbishop Ussher, naturalists such as Hutton and Lyell had observed the world around them and conferred on geologists more time to play with than they knew how to manage. Or had they?

By 1840 it was recognised that rocks were laid down, one on top of the other, in an orderly fashion. At any one time in the geological past particular groups of animals had existed, or a distinctive species dominated, and geologists used the fossils these creatures left behind as the means by which they could position a rock stratigraphically, relative to the younger one above it or the older one below it. In the Jurassic for example, along with all the dinosaurs, ammonites had evolved very quickly from one species to another, so ammonites are used to 'zone' the Jurassic, making it quite easy to accurately pin-point what part of the Jurassic a particular fossil is from. Eventually all the fossil-bearing rocks of Britain were arranged into a continuous sequential column, based on the fossils they contained, which was divided up into geological 'periods', frequently referred to as geological 'ages'.

A significant boundary within this geological column was recognised at the base of the Cambrian, a geological period named after the Latin word for Wales where these ancient rocks are typically exposed. The earliest fossils with hard shells date from the Cambrian period and it was during this time that the 'Cambrian Explosion' occurred as life on the planet became fully established, bursting into an enormous variety of weird and wonderful forms. This 'Base Cambrian' boundary was often interpreted by early geologists as the time at which the Earth had cooled sufficiently from a molten state for life on Earth to begin because, below the Cambrian, Precambrian rocks were thought to be almost completely devoid of fossils and frequently appeared to have been solidified from molten material. Now, however, we know that many Precambrian rocks contain soft-bodied fossils, clearly indicating that life existed long before the Cambrian, and that the Precambrian represents a period of time eight times as long as all the other periods put together.

Overlying the Cambrian the 'Ordovician' period is followed by the 'Silurian', both named after ancient Welsh tribes. The Ordovices were the inhabitants of North Wales and Anglesey who resisted the advancing Roman army until they were almost wiped out, and the Silures were the inhabitants of the Welsh borderlands. These rocks, widely exhibited in these parts of Wales, are renowned for the trilobites they contained – enormous woodlice-like creatures who scavenged the sea bottom and frequently shed their skins, leaving us to match body with head, millions of years later. Trilobites were a fantastically successful group, spreading in their different forms all round the world.

Above the Silurian lie Old Red Sandstone rocks, which, as the name describes, are represented by an enormous thickness of red sands and muds deposited in arid, desert-like conditions. The seas that lapped against this desert were first investigated in detail where they occur in Devon, southern England, and so this geological age is called the 'Devonian'. At the end of the Devonian the sea invaded the desert from the south, gradually spreading northwards over what is now Britain. Lush vegetation proliferated and in river deltas where land met sea huge mangrove-like swamps developed. Trees shed their leaves and finally died there, the debris building up year upon year, eventually forming the thick coal seams we see today in the 'Carboniferous' period.

At the end of the Carboniferous desert conditions formed again and prevailed throughout the 'Permian' and 'Triassic', this time depositing the 'New' Red Sandstone, typified by huge fossilised sand dunes. However, sand dunes are almost devoid of fossils, so are very difficult to position accurately in the geological column. At the end of the Permian the most massive of all mass extinctions occurred, with almost 95 per cent of all species disappearing in about 2 million years. It was the end for the trilobites, and most marine life was severely decimated. On land proto-mammals, amphibians and reptiles disappeared

Eon	Era	Geological Period or 'Age'	Age in millions of years	Life events
Phanerozoic	Cenozoic	Recent	0.01	Wide diversity of life
		Quaternary	1.8	Humanoids
		Tertiary	65	Age of mammals
	Mesozoic	Major extinction of life		
		Cretaceous	146	Wide diversity of life
		Jurassic	208	First birds and mammals
		Triassic	245	First dinosaurs
	Palaeozoic	Major extinction of life		
		Permian	290	Mammal-like reptiles
		Carboniferous	362	First reptiles
		Devonian	418	First amphibians
		Silurian	443	First land plants
		Ordovician	490	First fish
		Cambrian	544	First shells
Proterozoic Archaean Hadean	Precambrian	Base Cambrian Boundary		
			2500	Multi-celled organisms
			3800	Single celled organisms
			4554	Origin of the Earth

The geological column.
The geological column shows the geological periods or 'ages' to which fossils and the rocks that contain them can be assigned.

en masse. Life hung by a thread and took twenty million years to recover, but by latest Triassic times the first mammals had evolved.

In the warm tropical seas of the 'Jurassic', named after the Jura mountains in France and Switzerland, fish-life abounded in the sea, ammonites came in all shapes, sizes and colours and on the land dinosaurs ruled the world – it was a time of wide

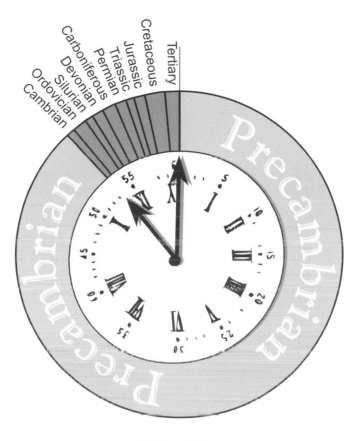

Geological time.
To give an idea of the enormity of Precambrian time, the table
opposite can be displayed as a clock where the Precambrian is seen
to occupy almost ninety per cent of geological time. Man appears a
few seconds before midnight.

biological diversity. It was succeeded by the 'Cretaceous' which
takes its name from the Latin word for chalk – *creta* – its most
conspicuous and widespread deposit, exemplified in the White
Cliffs of Dover. Chalk is composed chiefly of coccoliths, the
microscopic shelly remains of single-celled algae, and hence is
primarily of plant origin. It is impossible to imagine how many
must have died to make up one stick of chalk, let alone a whole

cliff-full. The Cretaceous seas covered a wide area of Europe and parts of central North America such as Kansas. Another major extinction of life forms occurred at the end of the Cretaceous, wiping out the dinosaurs and many other species. Whether it was due to the fashionable theory of a massive meteorite impact, or to a period of violent and prolonged volcanic activity that caused ash in the stratosphere to block out the sun, or a combination of the two, is unclear, but in many places around the world there is certainly evidence for tremendous volcanic activity at that time.

Finally, at the top of the geological column, the 'Tertiary' is the collective name given to all those rocks deposited since the Chalk and before the Ice Age. The Tertiary is frequently referred to by geologists as 'the gardening on the top' since, to a geologist, the rocks are so young – the Tertiary started 65 million years ago! Anything since the Ice Age is called 'Recent' – which seems to be self-explanatory.

But it was all very well having a geological column into which all the fossiliferous rocks of the world could be slotted when in fact only about fifteen per cent of rocks contain fossils anyway. The other eighty-five per cent are either sedimentary rocks devoid of fossils (although there are less of these today as we are able to identify algae and pollen spores with powerful microscopes), metamorphic rocks which are the result of rocks having been buried to great depths and pressures and so 'metamorphosed' into other rocks, or igneous rocks (from the Latin word 'ignis' meaning fire), which are the result of molten material that has forced its way up from great depths within the Earth and thus cannot possibly contain any fossils. The geological ages of metamorphic and igneous rocks are therefore often hard to assess and had to be guessed at according to their relationship with fossil-bearing rocks in the vicinity. But in 1840 even rocks that did contain fossils, and could therefore be assigned a geological 'age', still could not be 'dated'. In other words, nothing was known about the rock in terms of its absolute age in years, in the same way that you and I know how old we are in terms of years. This distinction between the geological age and the dated or 'absolute' age of a rock is an important one to bear in mind.

--------------------------- ❧ ---------------------------

Lyell tried not to be too specific about the age of any one geological event, but Darwin rushed head-long into attempting a calculation. In the first edition of *The Origin of Species* Darwin took as an example of the enormity of time required for geological processes, the erosion of the Weald, a large valley that today stretches between the North and South Downs in the south of England, but which at one time had had a great dome of rocks above it: 'It is an admirable lesson to stand on the North Downs and to look at the distant South Downs' Darwin wrote, 'one can safely picture to oneself the great dome of rocks which must have covered up the Weald within so limited a period as since the latter part of the Chalk formation'. Using rather sketchy figures for the amount of rock that had been present in the dome and making the assumption that the sea 'would eat into cliffs 500 feet in height at the rate of one inch in a century', Darwin boldly calculated that 'At this rate the denudation of the Weald must have required 306,662,400 years; or say three hundred million years.'

The implications of this calculation for the total age of the Earth were stupendous. If it had taken three hundred million years to erode the Weald dome down to the level we see it today, how much longer had it taken to deposit the Chalk in the first place before it could be eroded? And as the Chalk formation of the Weald was deposited near the top of the geological column, what did that say about the age for the miles of rock below it? There was indeed 'no prospect of an end'. It caused a furore, with one of Darwin's critics, James Croll, suggesting that he had been led astray by the geologists who provided him with vast time scales to bestow on evolution. The problem was, Croll believed, geologists were simply unable to imagine magnitudes as large as a million years, let alone countless millions of years. He suggested that once this 'deficiency' was recognised then Darwin and the geologists would see that geological processes were actually much faster than they realised, and hence geological time was not nearly as long as they thought.

Another critic was John Phillips, the nephew of William Smith. Phillips had undergone all his geological training alongside Smith and was now the Professor in Geology at Oxford University. He accused geologists of speaking with 'heedless freedom of the ages that have gone' and questioned whether their claim that the Earth was millions of years old was 'information of value'. He considered that this 'abuse of arithmetic' was likely to lead to 'a low estimate of the evidence in support of such random conclusions, and of the uncritical judgement which so readily accepts them.' Then, as perhaps now, geologists were often considered to be numerically illiterate: 'I dislike very much to consider any quantitative problem set by a geologist. In nearly every case the conditions given are much too vague for the matter to be in any sense satisfactory', complained the physicist John Perry.

Although Darwin later said that his calculation was only an attempt to give a crude idea of geological time, it was an attempt he was to regret. Immediately the calculation was seized upon and used as ammunition against the whole argument for evolution. Darwin's theory for the transmutation of one species into another was seriously discredited by this rather casual, and actually unimportant, calculation. A year later the American edition of his book carried a footnote revising some of the assumptions made and a confession that 'I have been rash and unguarded in the calculation'. By the third edition in 1861 it had disappeared altogether.

But worse was to come.

Darwin's Sorest Trouble

How immeasurable would be the advance of our science
could we but bring the chief events which it records into
some relation with a standard of time!

William Sollas

Like most small boys brought up in a religious community, as
would have existed in the Methodist enclaves of Gateshead at
the turn of the twentieth century, Arthur Holmes and his friend
Bob Lawson did not often find their favourite reading in the
Bible. But in years to come Arthur well remembered his parent's
Bible, and the magic fascination of the date of Creation, 4004
BC, which appeared in the margin of the first page. *I was
puzzled by the odd "4"*', he wrote. '*Why not a nice round 4000
years? And why such a recent date? And how could anyone
know?*' But all he learnt from his parents was that to question
the 'Word of God' was simply 'not done'. This Biblical time
barrier was further reinforced in Arthur's mind at Sunday School
through the teachings of Philip Gosse, a Victorian naturalist
who considered he had reconciled Hutton's geological findings
with the Scriptures in his distinguished book *Omphalos*. In this
work no compromise was called for. It was only necessary to
believe that the Earth was created about 6000 years ago, in strict
accordance with Biblical chronology, 'exactly as it would have
appeared at that moment of its history, if all the preceding eras
of its history had been real'. In other words, God had created

the world to look like it was old, just to fool all the geologists, and Nature was reduced to a role of meaningless illusion.

When Arthur and Bob went back to school for their final year, they excitedly told their teacher, Mr. McIntosh, about the discussion on radioactivity in *The Times* that was just coming to a close. It was Mr. McIntosh who had taught them to laugh at the teachings of Gosse and from whom they learnt with delight that there were people who audaciously doubted that the world was made in six days. It was he who introduced them to the works of Lord Kelvin, who had transformed four thousand years into millions, and now here was Kelvin being publicly derided in *The Times* by geologists who brazenly demanded more time! The boys' imaginations were fired. Both wanted to know about the magic of 'radioactivity' that they had read so much about, and were captivated with the thought of trying to determine the age of the Earth for themselves. But more urgent things were approaching and there was no time for day-dreaming. They took their Higher Certificate exams, passing at the top of their class, but Arthur did exceptionally well and was entered for a National Scholarship award which meant further exams. Eventually the long-awaited letter from the Board of Education arrived and no-one was more surprised than its recipient:

Sir, I am directed to state that you have been successful in the recent Competition for National Scholarships – Group B, (Physics). These scholarships are tenable at the Royal College of Science, London, or at the Royal College of Science, Dublin, and I am to request that you will at once inform the Board at which of the Institutions you desire to hold your scholarship.

He chose London, while Bob went as planned to Armstrong College, Newcastle. Both went to study physics.

At the Royal College of Science in 1907 there were only two terms a year. The hours of study were strictly between 10 a.m. and

Arthur Holmes aged about 21.

Bob Lawson, Arthur's best friend for many years.

5 p.m., with 1.15 to 2 p.m. allowed for refreshment, except on Wednesdays and Saturdays when study finished at 1 p.m. All students were expected to enter the times of their arrival and departure in a book kept in the hall, and any absence was immediately to be reported. Punctuality was a matter of great concern and late-comers would not be allowed into a lecture once it had commenced. Even the staff had to sign an attendance book, with those arriving after 9.50 signing below the red line. Smoking was strictly prohibited in any part of the building and students were required to refrain from shouting, whistling or singing. Where damage to equipment occurred as a result of carelessness, it was to be paid for by the student. 'Caution Money', a deposit of one pound per year, was required to cover such eventualities.

Dress and health were also matters for scrupulous attention. All members of the teaching staff were expected to wear dark blue suits with stiff white linen collars – the higher the better, some reaching as much as three inches. The Professor would never be seen outside the College on official business without a silk top hat, having first removed the dust by a quick polish on his sleeve. Students suffering from infectious diseases, or coming from homes where infectious diseases prevailed, would not be allowed into College, and amongst the information required on the admission form was a statement that the applicant was free from any organic disease or physical defect that would interfere with his studies. Life was formal and structured. It must have been very intimidating for a young boy from Gateshead away from home for the first time in his life.

The National Scholarship Arthur had been awarded was worth sixty pounds a year. It entitled the holder to thirty shillings a week during the academic year and free admission to lectures, laboratories and instruction during the four years necessary for completion of the Associateship course, equivalent to an honours degree today. At 1 p.m. on the first Wednesday following the opening of College, and at the same time on every succeeding fourth Wednesday, five pounds of the allowance would be paid to the National Scholars. Students failing to attend on that day would not be paid until the following month. However, even if they did attend, thirty shillings a week was barely enough to live on. A third class railway fare was allowed by the Board for one journey a year to and from London and the home of the National Scholar.

In the first year all science students, regardless of their intended specialisation, followed the same rigorous course of nine hours instruction in mathematics and mechanics, and twenty-one hours of chemistry and physics. The only rocks prospective geology students would see in their first year were those in the cobblestones as they rushed to their physics lectures each morning hoping not to be late. In the second year students had the option of taking geology with mechanics in addition to the subjects already studied, so it was during this year that Arthur Holmes was again exposed to geology.

Professor William Watts had joined the Royal College of Science only the year before Holmes. An accidental geologist, Watts had completed the geology paper in his exams at Cambridge instead of the impossible chemistry paper he should have been answering. As luck would have it both subjects happened to be on the same sheet. He grabbed the opportunity and over the next three hours the potential chemist became a geologist. Prior to his arrival at the Royal College, the teaching of geology had become a dry and dull affair, as witnessed by H. G. Wells who had once studied geology there, but failed his final exam. Wells acidly described the course as 'a great array of damn cold assorted facts, lifelessly arranged and presented. The exciting questions were never followed up – they were barely hinted at'. But Watts changed all that and by the time Holmes was studying geology there it was an exciting and dynamic subject. A man of great personality and charm, Watts excelled in the lecture theatre with his talks on elementary geology. Inspiring countless young men, and even some women, this was the course upon which the department concentrated its energies and which Watts took very seriously. The bell for the morning lecture rang at five to ten when Watts would be conducted to the lecture room by a member of staff. At precisely ten o'clock the lecture room door would be closed and locked, the register removed, and Watts would begin.

Watts often said it was his job to make and mould young geologists and he would strongly encourage them to read original papers published in the journal of the Geological Society, of which he was President. He also insisted that all members of staff should attend Society meetings on Wednesday afternoons in order that they keep up to date in their teaching. He once chided one of his assistants for missing a meeting, who responded that he had had a bad toothache, 'Well next time' replied Watts, 'try having your tooth out the night before'.

As a result of Watt's lectures, avid reading of geological material and his memories of Mr. McIntosh's lessons, Holmes began to think about becoming a geologist. Not only did he find the topic fascinating, but on looking around he felt that job opportunities for students of geology in the established mining industry and the emerging petroleum industry were going to be much greater than opportunities for physicists. Consequently, at the end of his second year Holmes took his BSc exams in physics, then a necessary step to be achieved before he could continue to the third year, and, somewhat to the consternation of his tutors in physics, changed over to study geology. In fact he never quite made the full transition.

In the same year that Holmes had started at the Royal College, a new Professor was appointed to the physics department. Professor Robert Strutt was a brilliant mind from the hallowed halls of the Cavendish Laboratory at Cambridge University where great names in physics were being made. Holmes was of course already familiar with the name of Strutt from the correspondence with Kelvin in the Times, and now here was the man himself. For someone who wanted to study radioactivity, Holmes could hardly believe his good fortune.

Like everyone else in the Cavendish at that time Strutt had been swept along with the rising tide of interest in radio-activity, and had been working on it when he became embroiled in the famous 'Times debate' with Kelvin. A year after the debate Lord Kelvin died, but he left behind him a legacy that the Earth was only twenty 20 million years old, which hung like a mill-stone around the necks of geologists, for the majority needed far greater amounts of time in which to shape the Earth. Strutt felt it was time to dispose of the millstone and was now working on the idea of using radioactivity to date the age of the Earth.

But how had it come about, this twenty million years?

In 1862 Lord Kelvin was the Professor in Natural Philosophy at Glasgow University and the world's expert on thermodynamics. A scientist of international repute and ferocious ability, he was then at the height of his powers and widely regarded by his con-temporaries as the greatest physicist of his day – a formidable opponent. In April of that year he opened his address to a meet-ing of the Edinburgh Royal Society with a blistering attack on geologists and their methods: 'For eighteen years it has pressed on my mind, that essential principles of Thermo-dynamics have been overlooked by geologists'. He went on to berate them for their insistence that the natural processes seen acting on the Earth today were the same as those in the geological past, and that therefore the rates of those processes had never changed over geological time. He condemned the geologists' unscientific demands for unlimited time and considered that the Earth had a very definite beginning and would also have a very conclusive end.

Kelvin argued that wherever you looked in the world, mines and boreholes showed that temperatures within the Earth increased with depth. From this he deduced that the Earth was still cooling down from a time when it had originally been a molten globe. Furthermore, he considered it was 'obvious' that if the temperature at which rocks melted and the rate at which they cooled down was known, then it should be possible to calculate the time at which the Earth's crust had consolidated.

Recognising the difficulties in using what were essentially unknown values for his calculation Kelvin initially allowed very wide limits: 'I think we may with much probability say that the consolidation cannot have taken place less than twenty million years ago, or we should have more underground heat than we actually have, nor more that four hundred million years ago, or we should not have so much'. Given these very wide brackets the intrusive impact of this physicist into the geological community, despite his rude comments, was initially quite small. Four hundred million years for the age of the Earth was very acceptable to most geologists at the time, and although twenty million would be stretching things a bit, no-one took that very seriously. But the subject continued to fascinate Kelvin. A few years later, with some improved data to hand, he revised the figures used in his original calculations, which brought the age of the Earth down to a more definite one hundred million years.

Many geologists were less happy with this, but Kelvin backed his arguments with all the authority that numbers and calculations could give him, and no one was better qualified to understand the laws of thermodynamics. Not only that, his speech propounding his new calculation was littered with more derogatory remarks directed at geologists: 'A great reform in geological speculation seems now to have become necessary'; 'It [Hutton's Theory] is pervaded by confusion'; 'a complete misinterpretation of the physical laws'; 'it is quite certain that a great mistake has been made', and so on. Although incensed by this attack, most geologists were clearly intimidated by Kelvin's authority with figures and felt obliged to heed his arguments. While some like Darwin, who described Kelvin as his 'sorest trouble', still clung to an intuitive belief that the Earth was much older, the majority tried to reconcile what they observed with a hundred million years. It triggered an explosion of attempts to date the age of the Earth which, by the 1890s, had become a geological obsession. The Dating Game had started in earnest.

As Hutton and Lyell had shown, geological processes continue all over the world in an endless cycle of erosion and sedimentation, erosion and sedimentation. The process could be likened to a gigantic hour glass in which erosion and sedimentation continue unceasingly. Every time it rains water sinks into the soil and promotes the slow work of decay by loosening particles. Every frost shatters the rocks with its expanding wedges of freezing water. Large tree roots help by splitting the rocks beneath the soil; smaller rootlets penetrate the crevices near the surface creating material that is easily washed away by the rain. Wind lifts up the lighter particles and carries them far and wide. From high crags and cliffs sharp-cornered fragments of all sizes fall down under gravity to the slopes below. The rain gathers in runnels and washes the mud, sand and even gravel into rivers which transport these particles to lower levels, ultimately depositing them at the bottom of the sea where they accumulate, over very long periods of time, into very great thicknesses of sediment. Eventually the hour glass gets turned upside down, the bottom of the sea is uplifted into mountains, and the whole process starts all over again. It therefore seemed evident to geologists who played the Dating Game, that if the rates of these geological processes could be measured, then they could be used to estimate an age of the Earth. Two of these 'hour-glass' methods prevailed.

As well as carrying their visible load of mud and sand, rivers hold in solution an invisible load: dissolved salts such as those of calcium and sodium derived from decomposition of the rocks over which the rivers pass. As far back as 1715 Edmond Halley, the Astronomer Royal who discovered the famous comet, proposed determining the age of the world by a rather novel method, utilising this information. He assumed that originally the oceans did not contain any salt but were pure water which had become salty over time due to 'saline particles brought in by rivers'. He therefore considered that if one measured how much salt was in the oceans now and then measured the levels again in a few hundred years time, one could estimate from the rate of increase how long it had been since the oceans had not contained any salt at all, and thus obtain the age of the earth – or at least the age of the oceans. He lamented the fact that no one had thought of doing this in Greek or Roman times so that we would now know the answer!

But Halley's proposal that measurements should start immediately seems not to have been heeded for a further hundred and eighty years until John Joly, a Professor of Geology at Trinity College Dublin, became the leading advocate of this approach in 1897. Joly considered that when Kelvin's 'molten globe' had cooled sufficiently to allow water to condense, a primeval ocean formed that did not contain much, if any, sodium. Therefore, if he measured the amount of sodium in the sea today and compared it with the amount extracted from the land and deposited in the sea each year by rivers, then he should be able to deduce how many years it had taken to arrive at the sea's present level of sodium. This method was fraught with assumptions and difficulties, not least of which was the measuring of sodium in rivers. Nevertheless, Joly arrived at an age of 89 million years since the formation of the first ocean; a value then comfortably within Kelvin's age of the Earth.

The second method for evaluating the age of the Earth was epitomised by Samuel Haughton, an Irish geologist, who introduced the principle that 'the maximum thicknesses of the strata are proportional to the times of their formation'. In other words, the thicker the strata, the longer it had taken to form. Having calculated that sediments were deposited on the ocean floor at the rate of 'one foot in 8,616 years', he then applied his principle and worked out just how long it had taken to deposit the total thickness of rock in the world. The enormity of the figure he arrived at forced him to conclude, 'which I am by no means willing to do', that even if 'the manufacture of strata in geological times proceeded at ten times this rate . . . this gives for the whole duration of geological time a minimum of 200 millions of years'. Now we know that he would have been much closer to the truth with his original estimate of 2000 million years, but we can also sympathise with his caution in those Kelvin-days of a young Earth.

But imagine trying to estimate how much rock there is in the world – and without computers! Not only was that an enormously difficult task, but it was also impossible to measure the rates of erosion and deposition accurately since they were different in different places and at different times. The values given for rates at which sediments were deposited ranged from one foot in a hundred years to one foot in nine thousand years, so inevitably the ages calculated using these deposition rates produced a similarly broad range: from 3 million years for the total age of the Earth, to 2400 million years just to the Base Cambrian. In fact it was a method that could produce any age you liked depending on how the parameters were adjusted, and thus the majority of ages determined for the age of the Earth fell, fortuitously, within Kelvin's estimate of a hundred million years.

Then in 1893 the results were published of some experiments that had been done in America that allowed Kelvin to refine his calculations on the age of the Earth even further. These experiments involved the heating of various rock types in a laboratory to discover what temperatures they melted at. The results dramatically reduced the rather generous value of nearly 4000 °C that Kelvin had assumed for the melting point of rocks to an actual value of 1200 °C. This considerable drop in temperature from which his molten globe had to cool resulted in an equally dramatic drop in the age of the Earth.

As the century closed Kelvin's final written word on the subject was published, gloriously entitled: 'The age of the Earth as an abode fitted for life'. With these new data on the melting point of rocks he determined that the time when the surface of the Earth was cool enough, and the temperature of the sun hot enough 'to support some kind of vegetable and animal life' was between 40 and 20 million years ago, with Kelvin's personal preference being for the lower value. There was an uproar from the geologists and biologists as even the most 'time-conscious' of these could see that 20 million years was just not enough for geological processes to have shaped the world, or for life to have evolved to its current state of complexity. The two sides seemed irreconcilable.

Mysterious Rays

The true men of action in our time, those who
transform the world, are not the politicians and
statesmen, but the scientists.

W.H. Auden

Once a year scientists of all disciplines came together to parade
their most recent discoveries, their 'wild miracles'. Held in a different place each year, meetings of the British Association for
the Advancement of Science were a forum for hot debate on current controversies. Consequently, in the 1890s, as arguments
raged about the age of the Earth, most BAAS meetings had an
'Age' discussion, and the Liverpool Meeting in 1896 was no
exception. That year it was the turn of the biologists to defend
their corner and Professor Poulton did this with fierce opposition to Kelvin and his meagre twenty million years. Responding
to a challenge from Kelvin – 'the burden of proof [falls] upon
those who hold to the vaguely vast age derived from sedimentary geology' – Poulton brilliantly put the case for an ancient
planet, and suggested that for biological purposes the Earth
must be more than one thousand million years old. He argued
that in order for the oldest fossils then discovered to be as highly
evolved as they were seen to be, inordinate amounts of time
must have previously elapsed while life forms evolved to such a
complex stage of development.

Unfortunately, Kelvin was not listening because he was

attending the session held in the Physics and Mathematics Section where all the talk was of 'mysterious rays'. The previous year, 1895, William Röntgen, a German physicist, had observed a mysterious source of energy being emitted as invisible rays. When he placed the hand of his wife over a photographic plate and in the path of these rays, Röntgen was able to develop a remarkable photograph that showed the bones in her hand, surrounded by the shadow of her flesh. This extraordinary image was the first X-ray ever taken.

The following year Henri Becquerel, a French physicist, wondered whether there was any connection between the newly discovered X-rays and the reason why uranium glowed in the dark after it had been exposed to sunlight. He placed some uranium in a draw with a photographic plate covered with black paper. Sure enough, on removing the paper the plate was seen to be fogged, proving that uranium too emitted invisible rays which were at least capable of passing through the black paper. Radioactivity had been discovered, and the first step towards finding the true age of the Earth had been taken. Speculation at the BAAS meeting as to what these mysterious rays were centred upon 'some kind of light', but at that time even the best brains in the business were mystified. However, set scientists a good problem and someone will try to solve it.

Initially, Becquerel's discovery did not arouse much attention, overshadowed as it was by Röntgen's X-rays because of the medical possibilities which X-rays opened up. But working in Paris at that time was a newly married couple, Pierre and Marie Curie, both of whom were physicists. Following the birth of their first child in 1897 Marie needed a project to work on and decided to make a systematic investigation of Becquerel's mysterious 'uranium rays'. Progress was quick. Within a few days she had discovered that another element, thorium, gave out the same rays as uranium. Then she noticed that regardless of what uranium was combined with, the strength of the radiation being emitted depended only on one thing: the amount of uranium that was present. This was an extraordinary result, a wild miracle of the highest order, because from her observations Marie was forced to conclude that the rays being emitted from uranium and thorium were not the result of a chemical reaction, but came directly from the element itself. At the beginning of the twenty-first century, with our detailed understanding of the atom and nuclear physics, it is difficult for us to appreciate the tremendous importance of Marie's discovery, which ultimately led to the most profound changes in our understanding of science.

The history of the atom goes back nearly two and a half thousand years to ancient Greece, around 400 BC, when the Greek philosopher Democritus made a significant contribution to metaphysics with his atomic theory of the universe. According to him all things originated from a vortex of tiny, indivisible particles – atoms. Objects differed only in the shape, position and arrangement of these atoms: atoms of a liquid were smooth and round while atoms of a solid were jagged so that they could catch on to each other and hold fast. Democritus coined the word atom, which in Greek (*atomos*) means 'undivided' because, according to his theory, atoms could not be destroyed. Two and a half thousand years went by before the theory developed much further, then, at the very end of the eighteenth century, Antoine Lavoisier identified the existence of ninety-two different types of matter. These were the elements – oxygen, hydrogen, helium, carbon, nitrogen, uranium, lead, gold, silver, calcium, zinc, silica and so on and so on. The building blocks of which everything is made – you, me, your car, my house, our dog and the whole shebang.

But how could the elements be classified? What characteristics did they all have in common which would allow them to be ordered? This was the great question of the day. Well, they were all made of atoms, the smallest and most fundamental particle of matter which, it was thought, could not be subdivided. So in 1805 John Dalton suggested that each element should be recognised by its 'atomic' weight, which was the weight of one of its atoms, because the weight of each element was different from all the others.

So for almost a hundred years the idea that the atom was indivisible had become firmly entrenched in scientific thinking, and now here was Marie Curie with evidence from the elements uranium and thorium that the atom itself consisted of even smaller parts, because it was the elements themselves that were emitting these 'mysterious rays'. Her recognition of this radical concept became the foundation stone for the whole of nuclear physics – the 'how' of dropping the atom bomb, the 'how' of men on the moon. And although at this point we are a long way from geological concerns, it was also to become the fundamental 'how' of dating the age of the Earth.

As Marie's work progressed she searched around for a variety of rocks that contained uranium, and discovered pitchblende, which was far more 'active' than it should have been, considering the amount of uranium it contained. She speculated that a new element must be present. In fact, she discovered not one but two new elements, both of which were significantly more active than uranium. The first she called Polonium after her native Poland; the second she called Radium from the Latin word *radius* meaning 'ray'.

Radium was to become a crucial tool in early determinations of the age of the Earth, but before that could happen Marie would first have to prove its existence by extracting it from the rock and producing it for everyone to see. It was a mammoth task. Only the tiniest amounts of radium were present in pitchblende, seven tons contained about a gram, so at least two tons would be needed to produce even the smallest visible quantity. Pitchblende was also very expensive, an American newspaper of the time reported that 'a ton of pitchblende carries about $15^{1}/_{2}$ grains of radium . . . this quantity is at present estimated to be worth about $2,000'. Fortunately, it was discovered that large heaps of radioactive pitchblende waste lay discarded in the forest surrounding the Joachimsthal uranium mine in Bohemia. Tons of this material were shipped to Paris.

Working in a makeshift laboratory that a visitor once described as 'a cross between a stable and a potato shed', Marie and Pierre began the laborious work of separation and analysis. Processing twenty kilograms of raw material at a time in a bubbling cauldron that gave off noxious fumes, it took them four years to isolate a few grains of radium. For their pains, in 1903, Marie and Pierre Curie shared a Nobel Prize with Henri Becquerel 'in recognition of the extraordinary services they have rendered by their joint researches on the radiation phenomena'. Radioactivity was here to stay.

———————————————— ⛶ ————————————————

The ten years that straddled the turn of the twentieth century must have been some of the most thrilling times Science has ever seen. The excitement over radioactivity and X-rays triggered an explosion of activity in physics labs around the world as new discoveries piled up one after the other. The greatest research school in experimental physics at the time was the Cavendish Laboratory at Cambridge University. It was headed by a brilliant physicist, James Joseph Thomson, who had first studied there under Robert Strutt's father, Lord Rayleigh, and whom he had succeeded as head of the laboratory in 1884. Not only was Thomson important for his own discoveries, but perhaps more significant was the fact that he was an exceptional leader. His ability to encourage and nurture others resulted in the Cavendish becoming responsible for most of the important discoveries made at the time, and world famous for the outstanding quality of its science. Although dating the age of the Earth was not the first, or even last, priority of many of the scientists involved, four wild miracles followed each other in rapid succession, all contributing to making it possible.

When J.J. Thomson detected the electron at the Cavendish in

1897 the atom finally lost its status as a fundamental particle that could not be subdivided. At first Thomson thought that the atom consisted entirely of electrons, but now we know that it also has a nucleus and that the electrons orbit in a field around the nucleus. However, it is hardly surprising that at first the nucleus was overlooked because its size, when compared with the surrounding field of electrons, is as small as a bullet in a battlefield!

The next discovery was made by Ernest Rutherford who, having also studied radioactivity at the Cavendish in Cambridge, was then appointed Professor of Physics at McGill University in Montreal, Canada, at the remarkably young age of twenty-seven. In 1902, working there with his assistant Frederick Soddy, they astounded the scientific community with the announcement that one element could change into another. Incredibly, it appeared that in the process of emitting 'mysterious rays', completely new types of matter were created, the chemical and physical properties of which were quite distinct from the parent atom: radium became radon – a metal became a gas! The ancient alchemist's dream of turning one element into another was happening before their very eyes – the only difference being that it was happening quite spontaneously without any help from Man. Wild miracle indeed!

Suddenly radioactivity was all the rage and the 'decay' theory of the break-up of atoms was a topic of supreme interest not just to scientists, but to the world at large. Journalists besieged Rutherford's laboratory and wrote fabulous and fantastic stories. Even the more staid headlines of The New York Times proclaimed 'Alchemist's goal reached by Briton' (actually Rutherford was from New Zealand) and discussed the 'Romance of Radium'. Doctors wrote letters to Rutherford about 'a trial of the inhalation of radium gas as a cure for tuberculosis', and 'the interesting effects produced when radium is brought near the eye. In a darkened room a rested eye, even with the lid closed, receives a sensation of light, no doubt due to the radiation going through the eyeball and affecting the retina'. One shudders to think of the damage done to these poor individuals.

The third discovery concerned helium, the lighter-than-air-gas we are all familiar with in party balloons which fly off into the stratosphere when young children accidentally let them go. Helium was not discovered until 1868, when it was first seen in the sun during an eclipse, and it was not until 1895 that it was identified here on Earth. Given that helium is the second most abundant element in the universe (after hydrogen) it is somewhat surprising that it was discovered so late – uranium for example was found in 1789. Once it was found on this planet, however, it was soon noted that helium invariably occurred in rocks that also contained radioactive elements such as uranium and thorium. It was therefore not long before Rutherford and Soddy suggested that the presence of helium in rocks might be related to radioactive decay.

In 1903, Soddy left Rutherford's laboratory in Montreal to work with Sir William Ramsay at University College in London, so in order to avoid unnecessary duplication, Rutherford had outlined a scheme of future work for Soddy before he left and it was agreed that Rutherford would examine whether helium was produced from radium. But Rutherford spent that summer in England and heard that practically pure radium bromide was being sold by the altruistic Professor Giesel at the very low price of about one pound per milligram. Both Ramsay and he purchased some of this material. Rutherford recalled what happened next:

I remember well a visit I made to Soddy at University College on the day when Ramsay and he were to collect the emanation [radon] from about .20 milligrams of radium to test whether they could detect its spectrum. Soddy told me he would take this opportunity of noting whether any helium was released from the radium.

Dissolving the precious radium bromide in water, the three of them waited expectantly for a wild miracle. Sure enough, the small bubble of gas that rose up through the water was found to contain the specific characteristics of helium. Generously, Rutherford lent Soddy and Ramsay his piece of radium bromide so that they could confirm the experiment.

Remarkably, it seemed, as radium decayed not one but two gases were being produced – radon and helium. But there was more! Soddy soon realised that radon too was unstable and when it decayed, it also produced helium, and was also transformed to yet another new element. The miracle was getting wilder and wilder.

Ultimately they established that in the 'decay chain' that started with an unstable 'parent' atom of uranium, a 'daughter' atom of radium was produced and helium liberated; in turn that unstable radium atom decayed to its 'daughter' product radon, also creating helium in the process, and so on through fourteen stages until eventually eight atoms of helium had been discharged and a completely new stable element was formed that had started life, a very long time ago, as an atom of uranium. But where did the decay chain stop? What was that stable new element? Well, it was to be almost another decade before Arthur Holmes confirmed that, but in the meantime the coffin lid was slowly closing on Kelvin's twenty million years.

The final event in this particular wild voyage of discovery was made by Pierre Curie at the same time as Soddy was working on helium, and a couple of months before he and Marie were awarded their Nobel Prize in 1903. With his young assistant Albert Laborde, Pierre detected that the radium he and Marie had isolated was constantly releasing heat. As electrons were explosively emitted from the atom, energy was being given out in the form of heat: 'Every hour radium generates enough heat to melt its own weight in ice' they announced. Here at last was the evidence geologists had been waiting for. While Kelvin might still be right in that the Earth could have been cooling down from a time when it had been a molten globe, what he had not known was that at the same time as the Earth was cooling, radioactive elements within the Earth were generating

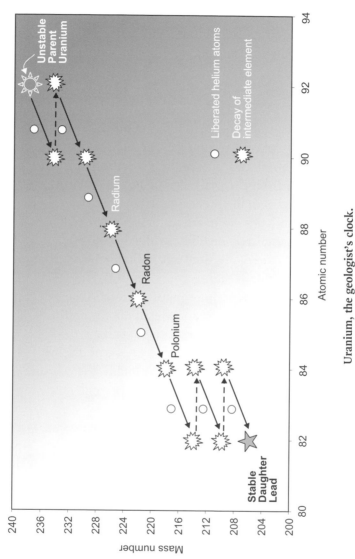

Uranium, the geologist's clock.

The decay chain of uranium (238) to lead (206) via radium, radon, polonium and other intermediate elements. Each explosive decay to the left is accompanied by the release of helium.

enough heat to prolong that cooling for longer than Kelvin could possibly imagine, and for as long as geologists and biologists might need it. The coffin lid banged shut.

And so it was uranium, as old as the Earth herself, that must surely be the Mother of Time, giving birth to the Daughters of Decay. Uranium was the clock geologists had been looking for which would allow them to tell geological time. So slow is her rate of decay, her ticking clock, that she requires four and a half billion years, almost the total age of the Earth, to be reduced by half her original amount, so today there is half the quantity of uranium left in the world as there was when the Earth first formed. In another four and a half billion years, half of what is here today will be gone. Of the daughters she conceives, radium will decay to half its original amount in 1600 years, a timescale we can begin to comprehend when we realise that 1600 years ago the Romans were leaving the shores of Britain after their four-hundred-year occupation. Radium leaves behind radon, which is highly unstable and intensely active. So rapidly does it give up energy and liberate helium that it decreases by half every four days. In its place arises another product of even more transitory nature, its activity falling by half every three minutes. And so on, continuing down a long succession of transformations, the shortest of which has a half life of less than a second! It hardly exists at all. But the most remarkable feature about all these radioactive phenomena is that they continue unceasingly: for year after year, decade after decade, century after century, millennium after millennium, the spontaneous production of helium goes on, accompanied by an unfailing evolution of heat.

At last the stage was set for geologists to escape from the physicists' strait-jacket of twenty million years. The clock had been found, all they had to do now was learn how to tell the time.

Doomsday Postponed

In radioactivity we have but a foretaste of a fountain of new knowledge, destined to overflow the boundaries of science.

Frederick Soddy

For scholarship students in London, college life was a permanent struggle against financial hardship. Sixty pounds a year was just not enough to survive on and Arthur Holmes' parents were not in a position to subsidise him financially. While he earned the occasional ten shillings reviewing books for *The Times*, the cost of living in London was a continual strain, and he was continuously on the lookout for ways of making money to fund himself and his studies. When halfway through the first year of his geology course Holmes saw an opening advertised for an 'assistant of the second class in the Department of Minerals' at the British Museum, he decided to apply and continue his studies part-time.

Appointments to permanent positions at the British Museum were then made by the three Principal Trustees of the Museum who were none other than the Archbishop of Canterbury, the Lord Chancellor and the speaker of the House of Commons. Holmes no doubt asked himself what on earth these somewhat inappropriate individuals knew about geology. A candidate had to be nominated by one of these Trustees, get on their 'list'

– Holmes was on the Archbishop's list – and then
undergo before the Civil Service Commissioners an examination in the
following subjects:
1. *English Composition*
2. *Translation from 3 out of the 4 following languages: Latin, French,*
   German and Greek
3. *Any other subject or subjects which the Trustees may prescribe, bearing*
   upon the work of the particular department in which the vacancy has
   occurred.

For the Department of Mineralogy nothing less than a knowledge of advanced mathematics, optical crystallography and
inorganic chemistry was expected, and in addition every candidate was required to 'satisfy the Civil Service Commissioner
that he is free from physical defect or disease . . . and that his
character is such as to qualify him for public employment.' All
this for a starting salary of £150 a year. The examination fee was
a hefty five pounds out of Holmes' meagre scholarship allowance, but since the salary would more than double his present
income he felt it was worth the investment of both time and
money.

The exams were known to be onerous and he spent from May
to October 1910 cramming for them, completely dropping his
college work. The results were finally published at the end of
November, and while Holmes came first in mineralogy – the
position, after all, for which he had applied – he came second
overall, Latin apparently letting him down. William Campbell
Smith was awarded the post and stayed there all his working
life. In years to come he was to present Arthur Holmes with the
Wollaston Medal, the highest accolade to be awarded by the
Geological Society. With hindsight Holmes must have been
grateful to William Campbell Smith for his greater proficiency
in Latin, but at the time it seemed as if all that work had been
a huge waste of time.

While waiting for the results of these exams, and having fallen
out of step with college work, Holmes was advised to start some

research. In their fourth and final year students were expected to contribute original research work for assessment with their final exams, so here was an opportune moment to start. As part of his physics lectures Professor Strutt had run a course on radioactivity and had been impressed with this young man's advanced understanding of the subject, so Strutt invited Holmes to help him with his research on radioactivity that he had started at the Cavendish.

Once Rutherford and Soddy had established that the helium found in rocks was due to the decay of radioactive elements, and that the heat generated by radioactivity was enough to keep the Earth hot for as long as geologists needed, it was but a short step to realising that if the rate of helium production could be established, and the amount of helium that had accumulated in a rock was measured, then a relatively simple calculation would show how long it had taken for the helium to accumulate, and the real age of the rock could be established. The difficult bit was in establishing the rate of helium production.

Do you by any chance remember the days before digital watches? When I was eleven years old my father gave me my first watch. In those days a watch was a symbol that you were growing up, it meant you were old enough to look after something expensive that was expected to last a long time, and I was immensely thrilled that my parents considered I had reached that age. But when I took the watch out of the box, I was a little disappointed. It looked very odd: in place of the usual numbers on the dial there were little green dots, and the hands were painted green as well. My father must have seen the look on my face because he told me to look at it under the bed clothes, and there I saw a wonderful sight. The dots and hands were glowing so that I could tell the time in the dark. I grew to love that watch as my most treasured possession and would place it each evening on my bedside table when I went to bed, so I could see it glowing in the dark. One night I thought to look at it under a magnifying glass, and was amazed at what I saw. The light from the dots was not steady but a quivering shower of sparks, like the tiniest firework imaginable. At the time I did not understand what it was I was seeing, but now I realise that each spark was generated by an atom of helium being emitted from the radioactive substance in the green luminous paint.

This same phenomenon was exploited when trying to determine the production rate of helium. An apparatus was arranged so that the helium atoms being emitted from a very small but very accurately known quantity of radium were impelled through a specially designed chamber. The passage of each particle set up a tiny electric current which gave a 'kick' to the needle of an electrometer. By counting the kicks, the particles themselves could be counted and the production rate of helium measured. This was, of course, an early version of the Geiger counter, now used to detect places contaminated by radioactivity.

But the decay rate was not all that Rutherford needed to know. In 1902 he and Soddy had determined the fundamental law of radioactivity. Written on a tablet of stone, the law said this: the number of radioactive atoms that decay in a given time, say a year, is dependent upon one thing, and one thing only – the number of radioactive atoms that are present. Put simply, the more atoms that are present, the more there are to decay. If there were a 100 atoms of a radioactive element in a rock, which decayed on average at 10 per cent a year, then after the first year 10 will have decayed, and 90 would be left, but after the second year only 9 will have decayed (10 per cent of 90) and 81 will be left. Thus the number that decay is directly proportional to the number that is present. Conversely, the numbers of daughter atoms increase in the same way and, provided that none of the parent or daughter atoms escape, then the sum of parent and daughter will always equal the original amount present. Thus when analysing a sample it is necessary to know not only how

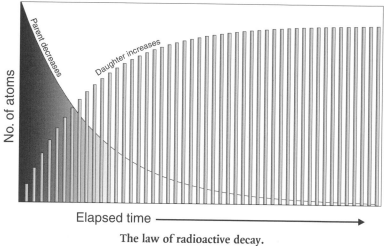

Parent decreases

Daughter increases

No. of atoms

Elapsed time ⟶

The law of radioactive decay.
The number of atoms that decay is wholly dependent upon
the number originally present. As the parent decreases, the
daughter increases in the same proportions.

much of the daughter element is present, but also how much of
the parent is left too.

By measuring the amount of radium and helium present in
his rock, and knowing the rate at which one decayed to the
other, Rutherford became the very first person ever to date the
true age of a rock. The age he obtained was five hundred million years. Five *hundred* million years! 'Doomsday postponed!',
proclaimed one newspaper reporting on this phenomenon. A
few weeks later, walking through the university grounds with
another small black rock in his hand, Rutherford met the
Professor of Geology; 'Adams', he said, 'how old is the earth
supposed to be?' The answer was the usual one prevailing at the
time – that various methods led to an estimate of 100 million
years. 'I know', said Rutherford quietly, 'that this piece of pitchblende is *seven* hundred million years old.'

Kelvin was not going to be pleased, but someone had to tell
him of this latest wild miracle. Unwittingly Rutherford drew the
short straw. He was about to give a lecture to a packed audience

at the Royal Institution in London when he realised that Kelvin was present. This apocryphal story is told in his own words:

I came into the room, which was half dark, and presently spotted Lord Kelvin in the audience and realised that I was in for trouble at the last part of my speech dealing with the age of the Earth, where my views conflicted with his. To my relief, Kelvin fell fast asleep, but as I came to the important point, I saw the old bird sit up, open an eye and cock a baleful glance at me! Then a sudden inspiration came, and I said, 'Lord Kelvin had limited the age of the Earth, provided no new source [of heat] was discovered. That prophetic utterance refers to what we are now considering tonight, radium!' Behold! the old boy beamed upon me.

But despite 'beaming' upon Rutherford, Kelvin was reluctant to relinquish his old ideas. He belonged to a school which believed that the energy expelled by radium was received from some external source and could therefore not be the cause of the Earth's internal heat. Although Rutherford's biographer A.S. Eve reports that 'With considerable courage, he [Kelvin] abandoned his theory publicly at the 1904 British Association Meeting', as we saw from the 1906 'Times debate' that so inspired Arthur and Bob, it appears that he never fully renounced his position.

Towards the end of his life Kelvin confided to a friend that he regarded his work on the age of the Earth as his most important contribution to science – greater, presumably, than his work on thermodynamics, for which he is now so famous, and for the successful laying of the first telegraph cable across the Atlantic, for which we should all be so grateful. I suspect he tremendously enjoyed all the controversy he caused, and while he may not have been right, he certainly made the geologists sit up and think about what they were doing, by making them question whether there really was 'no vestige of a beginning, no prospect of an end'. By proposing controversial hypotheses, Kelvin paved the way for a true numerical quantification of the age of the Earth.

But while Rutherford worked in Montreal, somewhat in isolation from the rest of the world, some duplication of his work inevitably occurred elsewhere. Strutt claimed that as soon as he read of Rutherford's discovery that helium was a product of radioactive decay, he realised at once that it could be used to measure geological time and had in fact done just that when Rutherford published his first 'age'. Like most other areas of life, it is the first person past the post who gets the prize and second place rarely gets a look in. But although Rutherford beat Strutt to the publication of results and so his name goes down in history as the first person to date a rock by radioactivity, it was Strutt who recognised the flaw in the method. Helium is a gas, and Strutt soon realised that much of the helium contained in the rocks was escaping into the atmosphere as the rocks were crushed up for analysis. This meant that only some of the helium produced by radioactive decay was being measured, and hence the age obtained was not truly representative of the age of the rock, it was only a *minimum* age. Strutt recognised that a new dating method was needed and encouraged Arthur Holmes to look for it. An aloof man with a retiring disposition, Strutt was nevertheless the perfect tutor for Holmes, leaving him very much to himself to figure out how to improve things.

In 1907 Bertram Boltwood, an American chemist, made a systematic analysis of rocks containing uranium. He noticed that along with helium, unusually large amounts of lead were present, and postulated that in fact it was lead that was the stable end product in the decay chain from uranium. If Boltwood was right, Holmes realised, then it should be possible to obtain an age by measuring the amount of lead present in the mineral, rather than the amount of helium. He decided to try. In the winter of 1910 he cut short his Christmas holidays in Gateshead, despite having to tear himself away from the comforts of home, his friendship with Bob and access to a good piano. Arriving back in London in the early hours of a cold, dark and rainy January morning, having been obliged by the conditions of his cheap ticket to leave Gateshead at midnight, he set up his apparatus in the cold, dark and quiet laboratory feeling homesick and lonely, but nevertheless excited at the prospect of starting work again.

Although the physics laboratory was new at the time of Professor Strutt's appointment, there had still been a serious shortage of apparatus which by 1908 had become acute:

We are at present largely subsisting on loaned apparatus, some of which belongs to other public bodies, such as the Royal Observatory, the Royal Society, etc., while some has been borrowed of private friends. I need hardly say that it seems rather below the dignity of an institution like the Imperial College that its teachers should have to beg apparatus of their personal friends for the purpose of teaching the students.

This appeal to the College authorities' dignity had the desired effect and elicited a grant of £700 for special equipment plus £800 for annually recurring expenditure which, although still not considered enough, did mean that by 1910 Holmes had access to some of the best equipment available at the time. He carefully unwrapped and wrapped it up, each time he used it.

The geologist in Holmes understood that rocks were made up of individual 'grains' of information called minerals, and that minerals held the key to understanding the origins of a rock because each mineral has a diagnostic colour, shape and hardness which result from the different elements it contains. To put it simply: elements combine to make minerals, and minerals combine to make rocks. So, when an igneous rock cools from its molten state within the Earth, different types of minerals will form depending on which elements are present in the melt. Thus when a mineral contains radioactive elements such as uranium and thorium, that mineral becomes a time-keeper for the rock it is in, because, as soon as the melt is cool enough for the minerals to crystallise, the radioactive daughter elements become trapped inside the mineral and the clock starts ticking.

Using the new uranium–lead technique, Holmes had decided to determine the age of a rock from Norway in which there were no less than seventeen different radioactive minerals, each one of which could, in theory, be used as an age-check on the others.

Holmes spent many tedious hours in the laboratory pains-takingly separating these from the crushed rocks. He then per-formed exquisitely delicate chemical preparations to isolate the required elements for measurement. His own account of the methods he used gives a good indication of just how laborious the work was.

First of all the amount of uranium present was measured, not directly, but by gauging the amount of radium emanation (radon) being emitted from a piece of radium. Having crushed up the mineral extracted from the rock, first in an iron mortar and then more finely in an agate one, the powder was *'fused with borax in a platinum crucible, and the resultant glass dissolved in dilute hydrochloric acid. After boiling and stand-ing for several days in a corked flask, the radium emanation was boiled out, collected in a gas-holder, and ultimately trans-ferred to an electroscope'*, which measured the amount of radon present.

While waiting for the radon to accumulate the lead was meas-ured by fusing the powder to a cake, whereupon it was broken up by boiling with water, twice dissolved with hydrochloric acid and twice evaporated to dryness. On the third addition of hydrochloric acid a clear solution remained from which lead was precipitated as a sulphide by heating and adding ammonium sulphide. The precipitate was collected on a small filter, dried, ignited, treated with nitric acid, boiled, treated with sulphuric acid and heated again. Eventually *'A tiny white precipitate then remained. This was collected on a very small filter . . . washed with alcohol, dried, ignited, and weighed with the greatest possible accuracy'*. Often the amount remaining was less than a tenth of a gram.

These chemical methods of separation required the patience of a saint, extraordinary dexterity and were incredibly time-consuming. Not only that, but in order to verify the results, each mineral analysis was repeated between two and five times, depending on how much material was available. At one point

Strutt made Holmes discard all the data and start again because radon had been leaking into the room, contaminating every-thing and giving spurious results (to say nothing of the effect it may have had on his health). Holmes had had to go cap in hand to the British Museum and ask for more sample. Frequently, he worked long and late into the night, but occasionally he went out to play.

Holmes' only real friend in London, Aquila Foster, was away in Germany studying geology in Freiburg, a city long famous for its teaching of geology. Arthur and Aquila had known each other from school days in Gateshead, but their friendship had not developed until they were thrown together in London. During the previous term Aquila's parents had moved from Gateshead to Palmers Green in London bringing Aquila's younger sisters and brothers with them. Having been a frequent guest of his friend's family while he and Aquila studied in London together, Arthur continued to visit them while Aquila was away. They were a warm and welcoming household, perhaps realising the need for this young man alone in London to have some sort of fam-ily life, so in the cold January of 1911 they gave a small tea party for his twenty-first birthday which helped him feel less home-sick on what should have been a very special day. The two older girls, Elsie and Edie, particularly entered his affections, and he confessed to his diary afterwards that *to make Edie my supreme favourite is no detriment to the others*. On being con-sulted by her parents as to whether Edie, then only fifteen, should stay on at school or not, Holmes admitted to having selfishly advised them that she should stay on at school and then go to college to study geology like her brother. In fact his moti-vation was the hope that she would eventually share his inter-ests. A home-loving boy, at the end of these weekends he often went back to college feeling sad and disconsolate at having to leave them. What made it even worse was that he was slowly sinking into debt. But help was at hand.

Memba Minerals Limited was a prospecting company that had recently purchased a licence to search for minerals of economic importance in Mozambique, then Portuguese-owned East Africa. A prospecting party had departed for Mozambique early in 1910, but having already spent a year there without any success, the company was now looking for extra recruits. The difficult terrain and large area to be examined had put the party behind schedule and Memba Minerals was nervous about getting its investment back. A former student of the Royal College was with the party and had sent back exciting letters about the hostile natives and the man-eating lions, so when further personnel were sought from the College, imaginations were stirred. In addition, the salary was rumoured to be £40 month, almost as much as a whole year's scholarship.

Interviews were arranged with Memba Minerals for three demonstrators in the geology department: Arthur Holmes, Edward Wayland and Alexander Wray. On writing home to tell his parents the exciting news Holmes was concerned they would be distressed at the idea of him going away to darkest Africa, but he was confident that they would not put any obstacles in his path *'for if successful, it will prove one of the best openings into the geological profession that I could have.'* How right he was may not have seemed apparent until some considerable time later.

While the candidates waited for the results of their interviews, Professor Watts gave his Presidential Address to the Geological Society. These events were prestigious occasions and everyone who was anyone in the geological world was there. So, keen to hear his Professor talk, Holmes inveigled an invitation from a colleague and went along although he was not yet a member of the Society. During the course of the afternoon Holmes met up with Dr Prior, Keeper of Minerals at the British Museum, who was keen that if the expedition to Mozambique went ahead they should bring back interesting rocks and minerals for the Museum as the area was completely unexplored geologically. At the end of the conversation, and perhaps anticipating the young man's potential, Prior invited Holmes to become a member of the Society and kindly offered to nominate him. Writhing with embarrassment because he did not have the necessary £5 fee, although dearly wishing to join this elite group, Holmes had to decline. *'Money will be the necessity. Influence I have in plenty for these Societies'*, he wrote in his diary that evening. It is events such as these upon which our lives turn. Holmes decided there and then that if offered a place on the Mozambique trip he would take it. Dating the age of the Earth would just have to wait.

The expedition's candidates were kept waiting for nearly a month before the results of their interviews were known, but after endless telephone calls all three were offered a six month contract as prospecting geologists in Mozambique, if they agreed to a revised salary of £35 a month each. There was little hesitation before they all accepted. The time passed in a rush of excitement. With less than a month before they were due to depart Holmes still had eleven laboratory analyses to be completed, an article on the results to be written, preparations to be made for the trip and farewells to be said to the family.

A month's advance on his salary meant he could pay off his debts and buy new outfits for Mozambique – two white 'safari' suites – as well as enough books to keep him occupied for six months and all the equipment he would need, including a pair of pistols in case of lions. It also meant that he could show off a little of his new found affluence so he treated his dear Edie with her younger sister Alice to the pantomime. Being held up on the bus on the way to collect them he leapt off it and grabbed a taxi, arriving in unaccustomed and impressive style. They

enjoyed 'Jack and the Beanstalk' enormously. Despite his extravagances he still managed to send £10 home to his parents.

A week before Holmes was due to leave, Aquila unexpectedly
returned from Germany. The friends had much to catch up on
and spent the night sitting over clay pipes and talking till two
in the morning. In later years Holmes was rarely seen without
his pipe, although he claimed not to smoke a great deal, *'but I
rather enjoy it. It rather mystifies the brain – at least in my
case, though it puts one in a good humour'*. Next day they both
departed for Gateshead by the steamship *Stephen Furness* so
Holmes could say farewell to family and relations. As the only
first-class passengers on the overnight voyage they had *'little to
do but read, play the piano, and converse with the stewardess
– the latter being Aquila's chief occupation!'*

The week in Gateshead was passed largely in music. Holmes
treated his mother to a piano recital – she enjoyed sharing his
new found prosperity in the expensive five shilling seats – and
his father to a concert in Newcastle. He and his friend Bob
bought some Grieg songs and practised them together which
gave them both *'the keenest of delight'*. Then it was time to do
the round of goodbyes. A large number of relatives and friends
were visited, including Mr McIntosh, the physics teacher at
school, who seemed impressed by Holmes' new job and must
have felt pleased with himself for having helped to turn out such
an accomplished young man.

A final evening was spent around the piano at Bob's house.
Amidst much teasing about Aquila's sisters, particularly one,
Holmes enjoyed being the centre of attention. But later, as he
walked home with his parents, he realised sadly that this might
be the last evening for some time in such enjoyable circum-
stances. At the station next morning many tears accompanied
the goodbyes, despite assurances that he was going to be alright
in Mozambique, and he was finally glad to be rapidly whirled
away, back to London. His last evening in England was spent at
the theatre with Edie watching *Henry VIII*, the tickets costing an

exorbitant ten shillings each. Feeling very grown up, they talked of the adventures that lay ahead of him as they dined at the Popular restaurant and taxied to Finsbury Park, just in time to catch the last train to Palmers Green where he stayed the night. Edie saw him off at Palmers Green station early next morning and promised to write.

In between all this Holmes somehow found time to finish the work in his laboratory. He wrote up the results whilst in Gateshead and, on the morning of his departure for Africa, he left them with Professor Strutt. It had taken months of hard work to complete, but finally it was all worth it. The rock from Norway had previously been assigned a Devonian geological age, and thus Holmes determined the first ever true 'date' for the Devonian to be 370 million years. It was a truly thrilling moment; his first very own wild miracle.

In addition to his own 'age' determination Holmes had also re-calculated some age data published by Boltwood, in line with more recent procedures, and had assigned a geological period to each of those ages, something which Boltwood had 'unfortunately omitted to do'. Boltwood was already well established in radioactivity circles, and it is unclear how he responded to this young upstart recalculating his data and ticking him off in print for not having ascribed a geological period to the results, but Boltwood was a chemist and not really interested in the geological aspects of his work. Holmes on the other hand, realised that there was no point in having an age for its own sake. To him the only thing that mattered was that each radiometric 'date' was a new point in geological time, so it had to be matched to a geological 'age'.

In fact Holmes' single age determination fitted perfectly with Boltwood's revised results, and Holmes laid out the radiometric ages alongside their respective geological ages so all could see that as the rocks became older geologically, so too did their radiometric dates. By doing this he was hoping to instil confidence in the radiometric methods, which were still treated

Geological period 1911	Holmes' ages (millions of years)	Geological period today	Age range today (millions of years)
Carboniferous	340	Lower Carboniferous	362–330
Devonian	370	Upper Devonian	380–362
Silurian or Ordovician	430	Silurian	443–418
Precambrian in:		Precambrian	
		Late Proterozoic	900–544
Sweden	1025	Middle Proterozoic	1600–900
	1270		
United States	1310	Middle Proterozoic	1600–900
	1435		
Ceylon	1640	Early Proterozoic	2500–1600

Holmes' first geological time scale.
The correspondence of Holmes' values with those of today is
remarkable, given the limitations of the techniques available to
him in 1911.

with considerable suspicion. One of the foremost American
geologists of the time, George Becker, had recently published
his *Age of the Earth*, concluding, after having reviewed all the evidence provided by the various hour glass methods, that the age
of the Earth 'must be between 70 and 55 million years', and that
consequently 'radioactive minerals cannot have the great ages
which have been attributed to them'. In similar vein John Joly,
in a serious attempt to reconcile radiometric ages with those
obtained from the sedimentary record, found it impossible to
accept the incredibly slow rates of deposition inferred by the
radioactive dates:

*If the recorded depths of sediments have taken 1400 million years to
collect, the average rate has been no more than one foot in 4000 years!
This seems incredible: and if we double the depth of maximum sedimentation it still remains incredible. But, if possible, still more incredible is*

the conclusion respecting solvent denudation to which radioactivity drives us. If the sodium in the ocean has taken 1400 million years to accumulate, the rivers are now bearing to the sea about 14 times the average percentage of the past. It seems quite impossible to find any explanation of such an increase.

It is difficult for us now to appreciate just how traumatic the discovery of radioactivity and all its implications must have been for the traditional geologists of the time. In a matter of a few years their whole world and conventional way of thinking about the Earth had been turned upside down. They had been given vast time scales to fill with sediments of which there was no evidence, so it was only natural that they should first assume a flaw in radioactivity, the young transgressor who had not yet shown his credentials. Quite reasonably Joly questioned some of the assumptions:

With these difficulties in view it is excusable to direct attention to the foundations of the radioactive method and ask how far they are secure. The fundamental assumption is that the parent radioactive substance, uranium, has always in the past disintegrated at the present rate. Is this assured? . . . I venture to suggest – I do so with diffidence – that our assumption of a constant rate of change for the parent substances – uranium and thorium – is really without any very strong basis.

Because, he argued,

we know nothing as to the origin of the primary radioactive elements . . . the rate of change 150 million years ago may have been many times what it is now.

Many geologists felt he had a point, although Holmes with his understanding of the laws of physics, never considered it for a moment.

Given the extremely crude methods that he had at his disposal, compared with the sophisticated techniques available today, Holmes' first attempt at a geological time scale has withstood the 'test of time' remarkably well. But in addition to his mineral analyses, Holmes' work confirmed Boltwood's ideas that lead was the ultimate decay product of uranium. Holmes clearly showed that as the ratio of lead to uranium increased, so too did the age of the rock. At one thousand, six hundred and forty million years old, the oldest mineral in the data set pushed Earth-time further back than it had ever been before. At the same time, an idea of the vast aeons represented by the Precambrian was beginning to emerge; already it spanned two thirds the total age of the Earth. With the new dating techniques, Holmes realised, it was going to be possible to impose an order on the undivided Precambrian rocks, something geologists had hitherto only dreamt of doing. At twenty-one young Arthur Holmes was starting to build order out of chaos.

There was only one possible flaw – what if some lead was already present when the uranium started to decay? If that turned out to be the case then, like the ages based on helium, his results would be meaningless, only this time they would be much too old. But that was a question that must await his return from Mozambique.

Holidays in Mozambique

A dusky maiden had a fit at the sight of our faces!!!

Arthur Holmes diary

At Victoria station on Saturday 18th March, 1911, Holmes met up with Wayland and Wray and a fourth member of the Mozambique prospecting team, a mining engineer called Wilson; the two leaders of the group, Reid and Starey, having gone ahead a couple of days previously. Despite a rough crossing from Dover to Calais the team were in high spirits and managed a fine dinner in Paris where the train stopped for four hours before continuing on through France, over the Alps and down the length of Italy to breakfast in Rome. They disembarked in Naples where they had arranged to meet Starey and were to board their ship.

At every stop on the journey to Naples Holmes had sent his parents a postcard 'to make the land part of my journey as real as possible to you'. None of the family had ever been abroad before, so even Europe seemed exotic and remote. As an only child Arthur was devoted to his parents, and they to him, and he was genuinely concerned for their worries about this trip to somewhere so distant as Mozambique. But he was an excellent and frequent letter writer, and almost everything that is known about this trip derives from the surviving letters he sent to his parents and to his friend Bob, as well as a diary that he kept for most of 1911. All the letters to his parents are rather formally, but nonetheless charmingly, addressed to 'My Dear Mother and Father' and end 'With love from your affectionate son, Arthur'. They provide us with a fascinating insight into attitudes of the times and the mind of a young man travelling abroad for the first time in the early part of the last century; a century in which so much changed so rapidly. The six months Holmes spent in Mozambique formed the foundation stone of his life-long research interests. In particular, his ideas of developing a geological time scale evolved during this trip, and so much of this part of the story is told in Holmes' own words.

At ten in the evening on Monday 20th March, 1911, they set sail for Mozambique aboard the General, a ship of the German East Africa Line which was on her maiden voyage.

Till about 7 p.m. we spent in seeing around Naples which is as different a town from anything English as could be imagined. The streets and roads are all paved on the same level with huge slabs of lava. Only the main streets are at all wide, the rest being so narrow that you could touch each wall standing in the middle. There are a few trams but mostly curiously shaped carts pulled by oxen with huge horns. All over there is a strong smell of garlic. Indeed all the cities we've been in seem to have a characteristic smell. Paris was like a bread poultice.

We left Naples on a small steamer which carried us across the bay to the great shining palace which was to be our home for three weeks: as we were going off there were dozens of beggars exposing all their infirmities – really most repulsive – in the hope of getting a few coppers. Having got on board, I found myself with Wray in the loveliest little bedroom imaginable. The ship can only be described as a first class hotel. There is a beautifully furnished drawing room with a grand piano. Dining room, smoking room, writing room and several others; library, map room etc are all magnificently fit up. There is a wireless installation on board and we get news every day from other ships. In the next boat to pass us Lord Kitchener is a passenger.

Of course on a boat like this it is only to be expected that most of the people will be exceptionally interesting. There is one family going to Nairobi, near Mombassa. The father is going big game hunting and the mother, three children with governess and servants are going to stay for a year in a bungalo in the hills at Nairobi. The eldest child is a girl of 13 called Dorothy – the nicest little girl imaginable. She is quite the favourite of the whole ship and I have made great friends with her. She is very like Edie except she is younger and has darker hair. There is a German young lady who is a professional violinist. We generally play together every morning. The rest of the people are explorers, missionaries, governors and consuls with their wives and children in some cases. We six are the only passengers to Mozambique. There are also two honeymoon couples who are going all round Africa as their introduction to married life.

Tips and suchlike here on board come to a quite considerable amount. I have had to borrow five pounds to meet my expenses.

Slowly the ship travelled across the Mediterranean, through the Suez Canal and down the Red Sea with the heat, even in March, becoming more and more intense:

All through the Red Sea the sun shone down furiously and at night one still felt cooked, for the air was hot and puffed like the fumes from an oven. One's intellect seems melted away and energy even to read a novel is not forthcoming. I spend a good deal of time on the grand piano . . . the heat paralyses all ones powers. It is quite impossible to think clearly. Everything is hazy and drowsy and even the atmosphere quivers with the strain. But the twilight abundantly makes up for this. Then all ones faculties become strangely intensified. The sun sets leaving behind sky and clouds of every colour, clear and limpid. The air is cool and refreshing and a strange sentimental influence seems at work. Unremembered incidents arise and cause a little homesick gulp and a tremendous feeling that the past was very good and of sorrow that it is past. This continues for an hour until darkness has spread its black garments over the sky. The warm luxurious climate makes me feel very sentimental and has influenced me, though not against my wishes, to write to Edie a confession of her position in my life. What she will think of it I don't quite know but absolute sincerity can do no harm and I hope she will return my comradeship.

He had to wait until September for her reply to reach him.

On Monday April 3rd the General and its passengers crossed the Equator but

the event was not kept in any way for a year or so ago when one of the passengers was going through the ordeal of being buried in soot and having the hose turned on etc., he became so nervous that he walked overboard and was drowned.

A week later they were at Mombassa and approaching the end of their journey:

As far as I can make out, the area seems to be rivalling the Riviera for popularity amongst the richer holiday makers. There is a railway to Lake Victoria Nyanza and half way towards the lake a new and flourishing holiday resort called Nairobi where half or more of the passengers are going. The climate is the finest in the world and already they have electric light, hotels, taxi cabs etc., and a constant service of trams to the lake and to the coast!

The British places, Mombassa and Zanzibar, are a curious mixture of native Arab and European houses and there are no direct streets, but narrow lanes which wind about like a gigantic maze, leading nowhere in particular, constantly devolving fresh views and surprises and characterised only by bad drainage and remarkable odours. All the people there in the hot, smelly streets, sleep out on their doorsteps. Last night at Zanzibar, Wayland and I went ashore after dinner and stayed till nearly midnight tramping through the narrow, devious pathways. There was a native wedding and when they saw us they played 'God Save the King' and shouted and danced.

The German places are totally different. They first of all blow up all the native kraals and then plan out a regular town like an English sea side resort – with a promenade along the sea front – hotels, gardens and so on. Apart from the palms and natives it is impossible to believe that one is in Africa.

I am very pleased with Africa so far and have been as much as four miles inland. Where we are going seems to be one of the healthiest points of the continent and we are all looking forward to a pleasant and interesting expedition.

Whether this was said for his parent's benefit or whether he was actually ignorant of the truth is unclear, but contrary to Holmes' enthusiasm for Mozambique, Reid, who had been exploring Africa for twenty years, had a very different opinion:

The territory as a whole is unhealthy, and the sanitary conditions in Mozambique and Mosuril are vile. The natives suffer much from leprosy and venereal diseases and in one or two places we noted entire villages had been deserted – from the native's description, I should say through the ravages of smallpox. Although the natives are excellent workers and will carry heavy loads for a 15 mile journey without a murmur, they are dirtier in their persons than the usual run of African native, and disgusting feeders, often never troubling to cook their food.

The capital town of Mozambique, also called Mozambique, was situated on a small island about seven miles offshore and was one of the oldest settlements on the coast of East Africa. The country had been occupied by the Portuguese for 400 years, but they had allowed it to become very run down and only recently had facilities such as the telegraph been installed. The expedition's headquarters were established in the town of Mosuril on the mainland, opposite the island of Mozambique.

Mosuril was in a state of ruin and endured only as a military post with a store, a post office and a few native huts, but the town of Mozambique did have a hospital and a cathedral.

Exactly three weeks after departing from Naples they arrived on April 13th, Maundy Thursday, at the town of Mozambique, where the party was met by Mr. Barton, a Rhodesian and the expedition's chief. After clearing customs they set off across the bay for Mosuril:

We went in a native sailing ship and with a good load of other luggage, several natives and eight of our party, the boat was heavily laden. It was the most exciting bit of travelling any of us had done. The wind was strong and drove the ships almost vertically and many a huge wave crashed in spray right over us giving us a thorough soaking. As we neared the shore we found it was high tide and that under those conditions the water completely submerges a small forest of trees. We thus had the unique experience of sailing over the tree tops! To land we were obliged to ride some distance on the shoulders of natives – Makua they are called – who waded very sure-footedly through the surf. We have a big Portuguese house, said to be 200 years old, surrounded by pleasant gardens and attended to by a large suite of Makua who make exceedingly good and willing servants – 'boys' as they are known.

Two years previously Memba Minerals Limited, a British company, had purchased mineral prospecting rights to a large tract of land in East Africa, then owned by the Portuguese. These rights had cost the company a hefty £105 000, and a further £25,827 7s 1d had been spent on the prospecting expedition of 1910 without them getting the smallest hint of anything worth exploiting. The only minerals which would have paid working were gold, copper and tin, and a few rarer ones they were unlikely to find. Iron was in abundance and were it in Europe would have made them millionaires, but in Mozambique it was value-less due to its distance from the market. It was therefore made quite clear to the team *'that we are being sent out as a last resource to the Memba Minerals Ltd. If we do not find any-thing, the Rhodesian Exploration Co. will not allow the Memba people any more funds and we shall be discharged.'* So the pressure on the team to be successful was considerable, but the conditions in which they had to work were daunting. Apart from the enormous size of the area to be covered – it was 200 miles just to the base camp at Sawa – the physical difficulties of get-ting around in this virtually unexplored country were over-whelming.

Had things gone as planned, the nine members of the group were to be split up into three teams which would separately make their way to the base camp at Sawa, the journey taking about two weeks walking fifteen miles a day, but set-backs started on day one. Huge numbers of 'boys', about a hundred and twenty, were required to accompany each team and had to be gathered together from the local Makua tribes. Sixty donkeys were supposed to carry the team members and much of the equipment, but when the donkeys arrived it was found that fifty-eight of them were diseased and had to be destroyed, resulting in each team needing a further eighty 'boys' to carry the extra loads, and everyone having to go on foot. While the first team of Cooper, Starey, and Wray managed to get away within a couple of days, it then proved almost impossible to round up enough 'boys' for the other teams:

In vain, attempts have been made in every direction to get natives as carriers, but the numbers we require are not forth-coming. The reason is that the Portuguese are going to have a war somewhere next year – quite out of our territory of course – and they are sending round press gangs obliging natives to recruit in the local regiments. Consequently the natives are much scared and they mistake us for the press gangs, and besides their numbers are necessarily reduced by the severe action on the part of the authorities. Thus at present we have nothing to do and are awaiting impatiently for more boys to free us from our imprisonment.

It was a further two weeks before the second team left, during which time Reid was severely bitten on the arm by one of the two remaining donkeys. He was sent to hospital the following day leaving Wilson and Wayland leaderless. Instead they had to take Tasker, the garrulous and eccentric camp organiser, who believed in reincarnation and that *'the ancients had some mode of overcoming gravity when they built the pyramids'*. He was going to be a great help in the field! They set out with only eighty 'boys', instead of the two hundred they really needed, but it was hoped Tasker would recruit more on the way.

With little to do until enough 'boys' arrived, and with concern mounting for Reid's arm, which they feared would have to be amputated, Holmes and Barton, now the only ones left at the camp, spent days in disappointingly familiar activities:

Monday: Today has been one of laziness and boredom. I commenced to read David Copperfield but felt a singular aversion to the high toned speeches which all Dickens' children indulge in. Tuesday: Very long weary day. Nothing to do and tired of doing it! Wednesday: Spent the morning in collecting butterflies. Bathed and walked to Conducia Bay. Dinner, bridge and to bed! Thursday: Spent almost whole day in tracing maps. A holiday of this sort was little anticipated by us as we came out.

But after a week of this boredom Holmes turned his thoughts to the work he had left behind. Although he knew that Strutt had approved the interpretation of his results from the laboratory, hastily thrust into Strutt's hands on the morning of his departure, he did not know how it had been received at the Royal Society where Strutt had read it on his behalf, after Holmes had left for Mozambique. He greatly feared there was something wrong with the method that made everything appear much older than it actually was, in the same way that the escape of helium had made all the ages analysed by that method appear much too young. Opposition to a great antiquity for the Earth was still prevalent amongst most geologists, who were firmly entrenched in their models of estimating an age by measuring either the rate that sediments accumulated, or the amount of salt in the oceans. The majority of results from these methods produced an age around one hundred million years which, following Kelvin's early lead, had become widely accepted as the age for the Earth – hence the outcry when Kelvin suddenly lowered it to twenty million – but equally problematic was that most geologists now seemed unable to adapt to suddenly having more than a billion years to play with. Where, for example, were all the extra sediments to come from to fill all that time? The possibility that one or two billion years might be available seemed to many geologists to be as embarrassing as the former limitation of twenty to forty million years. No, clearly there must be something wrong with this new method of radioactivity. It therefore must have been difficult for such a young man to stand against this strong tide of opposition without some feelings of intimidation, so instead of idling away the time in Mozambique, he set out to try and marry the ages obtained from the 'hourglass' methods – sedimentation and salinity – with those obtained from radiometric dates, by taking a new approach to estimating the total amount of sedimentary rocks present in the world.

Igneous rocks are the precursors of sedimentary rocks. They are the result of molten material that has forced its way up towards the surface from deep within the Earth. Originally the Earth's crust would have been comprised entirely of igneous rocks, but as soon as these were exposed at the surface they would have started to erode to form the first sediments. Consequently Holmes considered that a more realistic means of determining the age of the Earth by the 'hour-glass' method was to first deduce the amount of igneous rocks in the world from which sediments had been derived and then estimate the time it had taken to deposit the new sediments. *'Situated as I am in Africa without any geological literature,'* he wrote, *'I can only give the most approximate estimates'*. But he did not let that stop him! Taking his new approach to estimating rates of sedimentation, he calculated that it had taken 325 million years to deposit all the sediments that had accumulated since the Base Cambrian. He was thus comforted to find that this was a value much closer to the radiometric estimate of around 500 million years for the Base Cambrian, and felt relieved that radiometric ages were not as unreasonable as some geologists believed. In fact, because there really was no way of estimating the vast quantities of rock that had once been deposited but had since been eroded away over countless geological eras, this way of calculating the amount of sediments that had been deposited since the Base Cambrian was no more meaningful than any of the other hour-glass methods around, but the improved age agreement probably made him feel better.

He wrote a short article on his results and wrote to Bob asking him to see to its publication:

I want you to do me a favour which I am sure you will be glad to do. Read over the article and if you find it without fallacy, I should like you to verify my figures. Your correspondence with Dr. Jude is highly complimentary to you and is great evidence of the height to which you must have climbed mathematically. In that direction I feel absolutely ignorant – even though I know probably far more than the average geological man.

Another dig at the innumerate geologist!

Having sent my letter to 'Nature' keep your eye on the subsequent numbers and if it should be published get three copies. For the expense of the above 2/6 will probably suffice and I will tell mother to give it you.

He did not receive a reply from Bob confirming his acquiescence to this request until three months later, by which time it had already been published. It seems to have received very little attention.

———————————— ⊰⊱ ————————————

It was another month before Holmes was to go 'on safari' and life at Mosuril settled down to something of a monotony:

I like the life here very well. We are entirely a bachelor party and yet, through the services of the natives, we manage to get along very well without any of the fair sex! We generally arise about 6.30, being awakened by a servant bringing tea at 6. After a copious bath and shave (hair grows terrifically here) we breakfast at seven. Then commences the day's work. At 12.30 we have a four or five course lunch and are waited upon by three Somali boys arrayed in spotless white garments like shirts. One dish we always get is curry. After the first two days I managed to persuade myself that I liked it, and under pressure of hunger and the fact that it is one of the commonest camp foods, I have gradually evolved a liking for what was at first repulsive-looking green stuff.

Every afternoon we all go down to the sea and bathe. Then home again to afternoon tea! I am growing a huge tea drinker. The water alone is not very nice, having been boiled, so what we do is to drink tea to every meal, with coffee for a change at dinner. This we have at 7 in the evening and make it last a long time: sitting afterwards and listening to stories of the veldt and of hunting, told by the older men of our establishment.

It was around such a camp fire that he had his first lessons on 'investing' that were to stand him in such good stead, once he earned enough money to invest.

Occasional sorties into the hinterland were an excitement that broke up the routine:

On Friday I took a servant, well laden with our water bottles (filled with cold tea) and tins of fresh meat and bread etc., and set off about 6 a.m. to geologise in the district. I'm afraid

I'm being completely spoiled for I never by any chance carry anything – it would demoralise the natives not to allow them the honour! A personal 'boy' always goes along with one and looks after his master in everything. All that is to be carried is piled on to him, and the best of it is he really feels it an honour and so both parties are well pleased.

We walked about all day, and when we decided to lunch we found some cocoanut palms. Beneath these we had our meal, drinking the fresh delicious milk of young nuts. An admiring group of natives came around and took what was left. They sang in Makua, according to the translation of the boy, 'Great is the white man, who has come amongst us to visit us. Strong is his arm and beyond our understanding is his wisdom.' – and so time after time.

While based at Mosuril Holmes' ability to play the piano made him popular with the British and Portuguese residents on the island of Mozambique. He greatly enjoyed his visits there and the chance to make music. The town housed the residences of the Governor of Mozambique, several European businessmen and the fifteen staff of the Eastern Telegraph Company:

On Sunday I was invited to the Governor's with Allen from breakfast onwards. There we met all the big Europeans, the English and German Consuls, the Chancellor of the Exchequer of the colony, and several English from the Eastern Telegraph Company. One of the Englishmen is a Yorkshire man named Heselton who is a very fine baritone singer. The Governor's wife has a piano of sorts, and we had quite a lot of music which they appreciated immensely. I was quite a smart figure – dressed in white from head to foot even to shirt, tie and sash! I wish one could dress like that in England.

I have enclosed a small picture of Reid which occurs in the London Illustrated News of April 1st, 1911, and which we have just got. That will help to show you that the type of people I am with is one to be proud of and that there is nothing of the wild bush-ranger type of pioneer out here. In fact on

The Mozambique Expedition team.
Mr A.I. Reid seated far left and Mr E. Barton standing second
from left. This handsome group reassuringly show no indication
they were of the 'wild bush-ranger type of pioneer'.

*the coast, one has to dress very carefully and be exceedingly
polite, raising one's hat even to the Portuguese gentlemen.*

While in the town they took the opportunity to visit Reid,
who was still in hospital, now extremely ill with the pain in his
arm which had turned septic, and who was unlikely to return to
the party for some time. Barton was in a quandary as to whether
to proceed without him now that enough 'boys' had finally
been collected, or to wait until he was better. At last he decided
to head south on a short trip taking only Holmes with him,
rather than join the others at the base camp, as the cost of
keeping the 'boys' in camp doing nothing was becoming too
high. They started at noon on Friday 19th May, and Holmes
wrote excitedly in his diary: *'On Safari at last! With 40 boys
and 2 donkeys.'*

This Vegetable Prison

Feel rather homesick today – owing, I think, to the pervading smell of hyacinths here.

Arthur Holmes' diary

The route south that Holmes and Mr Barton followed had last been taken by Henry O'Neill in 1881, some thirty years previously. An adventurous British Consul, O'Neill was the first white man to penetrate beyond the coastal zone and although he lived on in the memory of some of the older natives, many of them, particularly the women, had never seen a white man before. Consequently Holmes and Barton were often followed by a shrieking mob for mile after mile. But this was the least of their difficulties. After a month of fruitless wanderings they stopped for a few days at Nacavalla where Holmes found time to write a letter to Bob explaining the problems:

The object of our expedition has been to find an old Arab Sultan, named Moravi, who, a quarter of a century ago was the ruler of the Makua over all the coastal district south of Mozambique. [The indigenous peoples of Mozambique are of Bantu origin, but by the tenth century the Arabs had established themselves on the coast.] *This man was attacked by the Portuguese, but instead of blotting him out they were themselves defeated. This however, was an unstable state of affairs and presently Moravi had to fly inland. He surrounded* himself with Makua chiefs on all sides and these have kept strangers from him all these years. We thought he might know something of the country's minerals – and further, because of his influence, his friendship would make the country open to us.

In reality we passed within a few miles of him, but the chiefs all said they had never heard of him, or else that he was far away and they did not know his whereabouts. Thus we got to Nacavalla, where we stayed a week, and at last impressed the chief there that he must let us through to Moravi. So at last Nacavalla [villages and their chief often had the same name] sent messengers to Moravi, 45 miles back towards the coast; and he, approving of us because we were English, sent us the necessary guides. Now we are back again at Nacavalla, having visited Moravi, and learned all he knows – disappointingly little.

All the country we have yet covered is singularly devoid of minerals and of gold there is not the faintest trace. Travelling is only very slow and difficult. At every chief's kraals one must stop and talk with him and tell him most confounded lies about being pleased to see him. When one arrives at a place it is imperative to stop – if you don't want to your carriers won't proceed, so what can be done? Then the chief comes and gives permission for the tents to be put up. You give him a present, about which he generally grumbles and threatens you with starvation unless it is increased! That is to say, he would tell his people not to sell food to the carriers for the cloth we give them every day. Cloth of course, takes the place of money here, and has the advantage that it can be used to cover up one's sex organs (!) but the disadvantage is that it wears out.

At many places the chief tells you he is going to have a dance in your honour. That is the promise of hell! The first time it is new and interesting, but afterwards it stimulates one to more bad language than a lifetime of golf (since golf

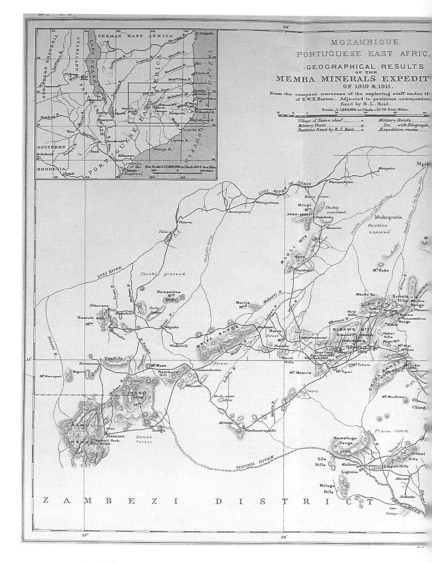

Results of the Memba Minerals Expedition of 1910 and 1911 in
Mozambique, Portuguese East Africa.

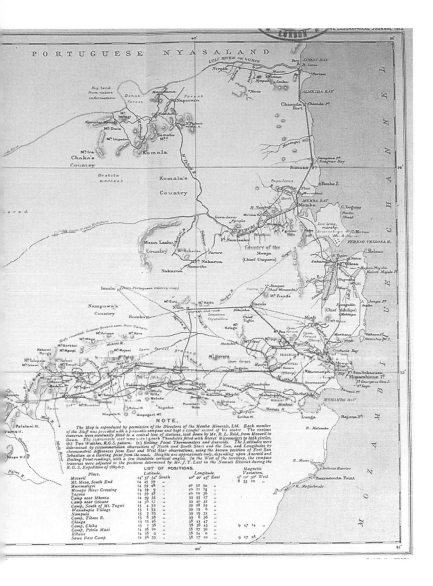

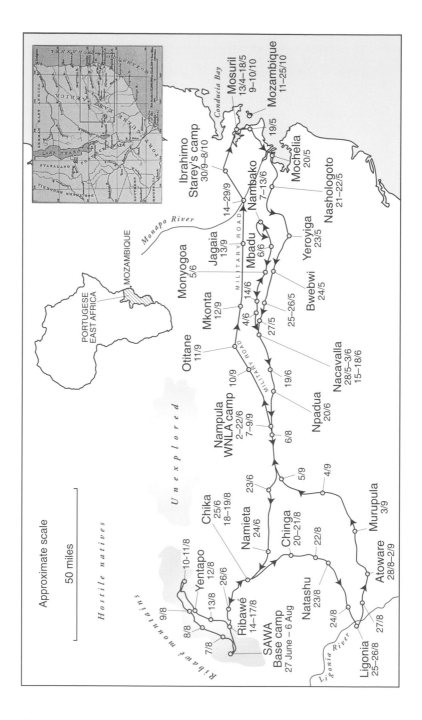

Approximate scale

50 miles

Hostile natives

Ribáwê mountains

Unexplored

PORTUGESE EAST AFRICA

MOZAMBIQUE

Monapo River

Ligonia River

8/8
9/8
7/8
Yentapo
12/8
10–11/8
13/8
26/6
Ribawé
14–17/8
SAWA
Base camp
27 June – 6 Aug
Chika
25/6
18–19/8
Namieta
24/6
23/6
Nampula
WNLA camp
2–22/6
7–9/9
Otitane
11/9
Mkonta
12/9
Monyogoa
5/6
Jagaia
13/9
14–29/9
Nambako
7–13/6
Ibrahimo
Starey's camp
30/9–8/10
19/5
Mosuril
13/4–18/5
9–10/10
Mozambique
11–25/10
Conducia Bay
Mochelia
20/5
Nashologoto
21–22/5
Yeroyiga
23/5
Bwebwi
24/5
Mbadu
6/6
14/6
4/6
27/5
25–26/5
Nacavalla
28/5–3/6
15–18/6
Npadua
20/6
19/6
6/8
5/9
4/9
MILITARY ROAD
MILITARY ROAD
Chinga
20–21/8
22/8
Natashu
23/8
24/8
27/8
Ligonia
25–26/8
Atoware
28/8–2/9
Murupula
3/9

84

seems to stimulate in that direction). First of all fires are lit and drums brought out and heated. Then a long bamboo stick is laid on the ground and the drums are arranged at the end. Then the stick and the drums are banged and the natives commence to yell and sing in the most indescribably hideous way until the din is simply terrific. The dancers are arrayed in beads, and around their ankles and wrists are large bracelets made out of dried fruits which rattle very loudly. Around their loins they have a belt with strings of these fruits hanging all around. The object of the dance is simply to balance on the toes and give a curious shaking movement to the abdomen which sets all the rattles going, and yet the appearance of the body as a whole is only that of a gentle movement forward as they slowly revolve in a ring. Really it can only be described as a very obscene dance, for they imitate all the actions of coition and the penis can often be seen to rise to an appalling size!

Map on opposite page

The route shown here is that taken by Mr Barton and Arthur Holmes while they 'geologised' in Mozambique. The dates indicate the places where they stayed the night. Each member of the expedition was charged with the task of keeping a careful record of his routes and was thus provided with a prismatic compass. The various traverses were eventually fitted to a central line of stations laid down by Mr R. L. Reid, from Mosuril to Sawa. The instruments used were: (a) 5-inch theodolite fitted with Reeves' micrometers to both circles, (b) two watches, R. G. S. pattern, (c) boiling point thermometers and aneroids. The latitudes were determined by circummeridian observations of North and South stars and the sun, the longitudes by chronometric difference from East and West Star observations, using the known position of Fort San Sebastian as a starting point for the coast. With the large party they had to take with them, and the difficulty of the terrain, the expedition would cover less than one and a half miles an hour. Nevertheless, in total Holmes must have traversed well over six hundred miles.

Head man and members of the Komala tribe.

Apart from the slow pace of the natives, the difficulty of the terrain and the need to stop at each village and haggle with the chief, a major problem that also held them up was sickness. As well as Reid, whose arm became progressively worse until he ended up having to go to Durban to get treatment, Barton was almost continuously ill with fever, frequently making it necessary to stay several days in one place while he recovered. Holmes was often left to cope on his own:

Yesterday I was in a great worry about him [Barton] *because he was so ill and the responsibility of a sick man on one's shoulders is really a serious matter. However, I got him along on a donkey for three hours alright, and then we arrived at a Portuguese station 10 miles away, and I managed to borrow a machila – a sort of bed carried on poles – and sent him on in advance in comparative comfort.*

The majority of the others also suffered some form of sickness. Wayland had fever for ten days after becoming lost in the forest when quite alone on an excursion from the camp at Sawa. He had been forced to lie out on the side of a precipitous river while it rained heavily all night. Another time *'Tasker was rendering the neighbourhood somewhat objectionable by his continual vomiting'*, while Cooper and Wray both had fever and were poisoned by bad bacon. Early morning starts were frequently delayed by finding that many 'boys' had gone down with sickness and fever during the night. The 'fever' was of course malaria, and within a short while of arriving at a place where the mosquitoes were sufficiently irritating to make Holmes put up his mosquito net for the first time, he too started having feverish attacks. He never mentioned this in letters to his parents to whom he was always in the *'best of health'* and the *'healthiest of the party'*, but as the fevers continued to get worse and more frequent he eventually confided in Bob:

Most of last week I was ill – not to be mentioned in the precincts of Low Fell, mind! [Where his parents lived.] *It wasn't at all bad, except that one felt so infernally weak. In fact, towards the end I began to like it – delicate soups for meals and port wine afterwards and strengthening drinks of milk, egg and brandy, and champagne to sleep on. And the joy of not having to get up at 5 a.m. in the cold dark, clammy morning – it was well worth having one's temperature up to 103!*

But I feel singularly devoid of news and interest today – perhaps because I've had a headache and taking quinine is beginning to have a depressing effect upon me. Lots of people can't take it at all as it makes them deaf and even blind for the time it is in their system – generally only three days. When we arrived here it was the end of the rainy season which had caused much malaria and blackwater fever, the latter being painfully abundant. Blackwater is simply complete physical break down as the result of months and months of fever. The name signifies that one's urine turns black, owing to the passing of blood.

While waiting around for Barton to get well, tribal chiefs to approve their route, or the recapture of donkeys that had run off, a favourite past time was hunting. This was not done for fun, in fact Holmes expressed an abhorrence of unnecessary slaughter *'it is sheer butchery to kill when there is no useful purpose to excuse it'*, but to supplement their food stocks and keep their diet as varied as possible. Apart from the occasional wildebeest they rarely caught anything, but nevertheless they did not do badly for food:

We have tea, coffee and cocoa and plenty of Nestlé's milk. Bread and cake we get made every few days and butter and jam we have in plenty. Our usual breakfast consists of porridge, and then eggs and toast. That we have in the dark while the boys are pulling down the tents and getting their loads ready for the day's tramp. We have plenty of tins of preserved and dried fruits and these we never miss for lunch and dinner. Generally we have a cold chicken at lunch and a grilled one for dinner, with soup made out of the leavings at lunch. So you see we fare pretty well. For the last few days however we have been bereft of fowls, and have been reduced to tinned sausages and sardines – of which one sickens very rapidly. Today we are in great luck, for our present from the chief includes a fine goat, which will keep us in meat for several days. This is a very welcome change. We also got some flour made out of mealies [maize], which makes very good porridge, and some 'beer'. The latter looked like sour milk and we immediately sent it off to the carriers.

Somewhat to Holmes' disappointment, there were few exciting encounters with wild life and little that really frightened him, apart from waking up one night to find a large snake in his tent. Leopards and lions, however, could be a problem, prowling round and stealing livestock from the camp neighbourhood at night: *'There was a dog here – a sort of mongrel pointer – quite a nice affectionate beast, and pleasant to have about the camp. Soon after going to bed the night became very quiet and a sudden series of piercing yells broke from the dog going close by the fire between our tents. There was a complicated shuffle and the dog flew for his life.'* It was never seen again, eaten, presumably, by a leopard. Another potential encounter with a leopard occurred as Holmes was exploring some caves which

showed clearly that they were the abodes of leopards and we went at it in grim excitement, pistol in one hand and candle in the other. However the usual inmates were absent and beyond shooting at a baboon we had no need for our weapons. On the whole, I must say that travel is less exciting than I anticipated and apart from that snake I have seen nothing from which an attack would mean death. Travelling here is as safe as in England, provided one uses discretion and does not shoot down occasional niggers as the Portuguese do.

Despite this bravado, they did pass four human skulls stuck up on posts at the entrance to one village.

After their rest at Nacavalla, Holmes and Barton finally left for the base camp at Sawa calling in *en route* to Nampula, which, a hundred miles inland, was the end of the telegraph line. Waiting for them there was a telegraph message with the good news that Reid's arm was much better and that he would be joining them from Durban in three weeks time. They then pressed on to Sawa. After the long trip of six weeks with only Barton for company – *'I sometimes feel very lonely with only him as a companion for we have no interests in common'* – the excitement of seeing his friends again at Sawa put Holmes in an elated frame of mind as he walked the last few miles to the base camp:

Oh, I was in a gorgeously happy mood – rather spoiled by Mr. Barton's illness, but all the same irrepressible – perfect 'joie de vivre'. It was a lovely morning, fresh and cool, and with a blue radiance in the atmosphere bathing one in a crystalline blue sky which thrilled one. All around great mountains in purple and gold were rising up to stupendous heights getting apparently even higher as we approached the wonderful valley. I sang out (misuse of 'sang' but you'll understand) all of the Greig's songs that I knew and simply revelled in the beauty of the morning. Then I got at last to the camp, where I was very thankful once again to meet some really good English fellows.

But perhaps the greatest joy of arriving somewhere like the base camp was the pleasure of reading the letters that would be waiting for him. They usually arrived about a month or five weeks after posting in England, but for some reason the first letters sent to Holmes at Mosuril did not arrive for eight weeks, causing him much anguish as he waited.

28th April: There was a mail came in yesterday and nearly everyone got letters except Wayland and I. It was a little disappointing, but I suppose the next boat will bring us news from home. 2nd May: Allan returned from Mozambique but brought me no letters – a matter of disappointment. 5th May: There was a mail in today – but still no letters! They take the devil of a time to come. Next week, surely, cannot fail to bring news.

At last some arrived on the 12th of May and he confessed to Bob how upset he had really been:

It gave me the most acute delight today to get your letter, with one from home and a Times. Having gone over eight weeks without having heard from anyone in England, I was beginning, rather foolishly, but nevertheless having the feeling, to despair. And every mail bringing piles of letters but never any for me gave me a pang of disappointment and irritability of manner which I found difficult to conceal. Now, having heard from you I feel quite wild with excitement, and have for the first time realised in a keen way, how far I have travelled.

So arrival at Sawa heralded not only the company of people his own age, but also the pleasure of more letters from home. That evening he sat down to reply to his parents and describe his new surroundings:

The Sawa valley is only one of five which intersect a vast extent of gneiss-covered granite intrusion mountains. [Gneiss is a rock that has been buried to such depths that the temperatures and pressures have altered (metamorphosed) its original structure and mineralogy – hence it is a 'metamorphic' rock. If the heating is sufficiently great the rocks may melt to form granites, consequently the two are often found in association with each other.] *The height of the camp is 2,000 feet and the mountains rise up with appalling swiftness to 4,000 5,000. Proceeding westward from the coast, the country begins to be diversified with isolated peaks or clusters of hills, which rise abruptly from the surface to the plateau and exhibit the most remarkable outlines, varying from gracefully rounded domes of smooth and naked gneiss to irregular knobs and pinnacles. In Mozambique the mountains are everywhere held in great veneration by the natives, and in the west this reverence for Nature's architecture amounts almost to mountain worship, for the Lomwe people have a legend that the first man and woman were born from the white peak of Namuli.*

The valley [at Sawa] *rises steeply but the river flows out into the everlasting plateau of this part of Africa – and a few miles across an imposing spectacle is afforded by another mighty range of abrupt peaks. These mountains attain a sufficient altitude above sea-level to ensure a fairly continuous rainfall throughout the year, and their slopes are therefore well watered and luxuriantly clothed with vegetation. Above the almost impassable jungle formed by the intertwining of vines and creepers rise groups of towering trees, whose interlocking branches provide a cool canopy in which innumerable hordes of monkeys find a congenial home. Here and there the traveller emerges into open glades of park land, which offer a charming contrast to the monotony of the forests, and on the banks of the watercourses an abundant flora delights the eye, dappling the fresh greenery with glowing clusters of delicate flowers.*

The country near the coast is characterised by palm and

A native hut at the Sawa base camp, behind which a granite peak
rises with impressive steepness.

*grass, the latter often being twelve feet high. Further inland
the palms disappear except for the banana, and the whole
country is a plain of deadly monotony. It is covered by a thick*

vegetation, absolutely impenetrable except along the six inch paths which the natives use. These are fairly good underfoot but needlessly long, passing round every obstruction and winding away from the proper direction, in fact following the footsteps of the first man who found his way through the bush from one point to another. There is a large traffic along these narrow paths – boys carrying monkey nuts to the coast. One gets very weary going through this vegetable prison day after day.

The camp itself is a great success. There are two big store houses, one house for Barton, another for a hospital, and two living houses, kitchen, garden by the river, and so on. For meals we have a sort of trellis-work summer house built on a little hill which commands a view of the whole camp and away out of the valley mouth, over the plateau beyond and to the precipitous heights on the horizon.

The garden he mentions was not ornamental but a way of supplementing their rather restricted diet, by growing their own vegetables.

Struck by the beauty and wonder of it all he wrote to Bob next day:

The stars here are wonderfully impressive. The Milky Way stretches across the heavens and gives one an unbelievable feeling of awe. Last night looking out of the door of my hut I had an indescribable emotion of nervousness – not of anything tangible, but of the overpowering majesty of the hills and of the stars and the wind playing infinite chords on the strings of the forest, and the calls of animals sometimes sharp and shrill in a death struggle, sometimes the roar which is born of a full stomach. I felt somehow what a fearful meaningless tragedy the whole Universe appeared to be. Have you yet seen Dr. Russell Wallace's newest book 'The World of Life'? It is good while it remains scientific but philosophically and imaginatively it is insanely absurd – the sort of book which ministers will rave over – excusing God's ways to Man and pointing out the contemptibly conceited idea, that the whole purpose of existence is MAN. You might read it during the vac. It is stimulating, but rather disappointing when he harps on the 'Purpose of God'.

Christianity makes no progress on the African coast. Mohommedanism holds almost entire sway and I honestly feel it is a better religion altogether for a black. It has the advantage of including Christianity, for Christ is one of the prophets and is considered second only to Mahomet. The boys say Christ was never married and so cannot afford to be a guide as to how wives and children should be treated – not a bad criticism and very original!

After a month at Sawa during which time Holmes compiled maps of the area, tested minerals and collated information provided by the team for a report to the company, Starey one day brought in a mineral that no one recognised. Apparatus available in the field for the identification of minerals was limited to the use of a blow pipe, which was simple, but effective. Powdered samples of the mineral under investigation were examined by heating them on a charcoal block, either in their pure state, or mixed with various other compounds. Air was then directed at the flame via the blow pipe through which is was necessary to breathe both in and out simultaneously in order to maintain a steady flame, whilst retaining a reservoir of air in the cheeks! The resulting reactions or colours seen in the flame being diagnostic of the mineral under examination.

After various tests Holmes proclaimed that the mineral found by Starey was none other than thorium, one of the radioactive minerals that he had been using for his dating work, which caused much excitement. By coincidence the post on that day brought a letter from Professor Strutt congratulating Holmes on the reception of his paper at the Royal Society meeting, and including twelve copies, now that it had been published. Holmes graciously distributed these around the members of the team, and that evening wrote excitedly to his parents:

Holmes' tent comprised a 60 lb load for a native 'boy'. They
would carry this weight for 15 miles at a time.

*This work of mine is, I think, going to pull our company out
of the mud of bad fortune for I have recognised minerals
here which none of the others had any idea of, and which
are the only valuable things we have found. I am getting rather
an authority out here! My paper, of course stamps me as a
specialist in that work.*

*I am going out shortly with Barton and partly by myself to
explore a range of mountains [Ribawé] for Radium minerals.
I am greatly looking forward to this piece of independent work
and organisation, and hope to make it a success. I shall have
about 12 or 15 carriers, a personal boy, a donkey boy and a
cook. My own things usually fill up four big boxes, but I am
only taking one with me because of the short time. That makes
one load. My tent makes a load; table and chair and bedding
another; trade goods and cloth, beads and bracelets and so
on, together with my bed are the fourth; a box of food the fifth;
tools and pots and pans and pails etc the sixth. My cook and*

personal boy also have bedding and that goes with one of the lighter boxes. Sixty pounds each is the most that a native will carry.

I have a jolly little personal boy just now. My others having been rotten, I sacked them. This one is called Morweha which means 'Bring it here'! I am greatly looking forward to making this little trip on my own. I am only going about 30 miles from here and shall work in the district for a week and then move south until Mr. Barton lets me know further plans. I shall be too busy to be lonely, and any how I have a boy Ali who speaks very good English. I can also speak Makua fairly readily now – tho' I find it difficult to understand, they talk at such a prodigious pace.

Holmes greatly enjoyed this first opportunity to work on his own. The district contained ancient Precambrian rocks that had never previously been examined, and his work went a long way towards determining the secrets of the geology of this extraordinary country. He collected samples of the different rock types, some of which the natives crushed in a large iron mortar. He then panned the crushed material in the hope of finding minerals of economic worth, particularly radioactive ones. But the search for radium and thorium proved as elusive as the search for any other minerals of economic importance, and having explored the river where Starey had found the original sample Holmes discovered *'to my disappointment that my thorite is a mixture of hypersthene and tourmaline'*, two common minerals of no economic value. Alas, he was not going to save Memba Minerals after all. He does not record, however, how this news was received back at the base camp!

Although unsuccessful in finding radium, the heavy residue left in the panning bowl often contained zircons, a mineral ideal for age determinations. These were sent back to the lab and later analysed by Holmes, allowing him to understand the age relationships of these ancient rocks not only to each other, but to Precambrian rocks around the world. The six months of geolo-

Sketch map drawn by Holmes in a letter to Bob Lawson showing
where he 'geologised' on his own in the wilds of Mozambique.

gising in the wilds of Mozambique, particularly the time spent
on his own, led to a life-long interest in the Precambrian, which
he spent a great deal of time trying to unravel.

When Barton caught up with Holmes they spent another
month geologising as they gradually worked their way back to

Nampula *en route* to the coast, the camp at Sawa having been broken up now the work there was completed. When they arrived at Nampula they were forced to stay there for a couple of days after many of the 'boys' who were from that region deserted, but that had its compensations: *'I had quite a big budget of letters including ones from Elsie, Edie and Mrs Foster. Edie retains her characteristic coolness and rather says 'consider yourself squashed'. However she appears to be good chums and causes no embarrassment whatever.'* He seems to have forgotten all about his feelings for her of a few months ago!

While waiting at Nampula, Barton unexpectedly received a telegraph from Memba Minerals in London, but as it was in code and his code book had gone ahead with the luggage he could not read it immediately. Sending telegrams in code was common practice since anyone along the long line of telegraph operators could have read the message. It was a few days before they caught up with their luggage, found the code book and deciphered the message. It contained a considerable surprise.

Thursday 14th September, 1911. My Dear Bob: I am greatly excited tonight and you will be too when I startle you by the unexpected news that as you read this I shall be on the ocean blue sailing homewards. Mr. Barton had a cable from the London office, but as he had not his code book with him, he was unable to read it until today on arrival here [camp at Monapo River]. *It was to the effect that everyone will leave Mozambique at the end of the agreement period, October 12th, a sudden change of plan which he is unable to understand. As there is a boat (I think) leaving on October 14th we shall of course, come by that, arriving at Southampton on or about Nov. 20th. Hurrah! I should have liked very much to have stayed the other month for the sake of the extra money, but I must say, the prospect of being home so much earlier rather fills me with delightful excitement.*

After a week's glorifying at home I'll return to London for a fortnight to start myself at College which I mean joining again from November 20th – I must get in as much time as possible to finish my Diploma course, in case I have to leave on another job. I'm awfully keen to be back at work as I have heaps to do. Besides a probable paper in conjunction with Wayland on the geology of this 'ere spot, I shall work on an independent paper as I have collected some fine rocks and come across some very interesting geological complexes. I intend joining the Geological Society and the Royal Geographical Society and want the candidatures ready for the new year's membership. Lately I have been full of ideas for research work and shall be very keen to get started again.

One thing which has struck me in connection with earth cooling and the condition of the earth's interior is this – and I want to discuss it with you on my return and ask your help in the maths involved: You know that the radium in the earth's crust is easily sufficient to account for the radiation of heat into space. It would be interesting to see what kind of temperature gradient the corresponding amount of radium of 1500 million years ago would give. There was sea then and therefore the temperature of the surface [of the Earth] must have been under 100 °C [otherwise the sea would have boiled away]. From rough calculations I have made out here (necessarily not very reliable!) I estimate the Radium at 20% more [than today] and the temperature at 60 °C – assuming there is exactly enough radium for the present radiation. If that be the case under rigid maths, it almost proves that the Earth is cold in its interior as otherwise the earth would have notably grown hotter.

A big problem with the discovery that radioactive elements kept the Earth's interior hot was that if, as it was first assumed, the elements were evenly distributed throughout the interior then they would be generating such heat that, far from cooling down, the Earth should be heating up. So Holmes had the clever idea of turning around Kelvin's method of calculating the age

> I shall now be able to write my article on
> the age of the Earth's crust before Xmas & get it
> in the Feb. no of Science Progress with luck.
> I intend writing all over the world to surveys
> & societies for material of known geol. age
> to analyse for U. & Pb. I am in hopes
> of gradually building up a geol. time
> scale & hope it might do for a D.Sc. !!!
> There's conceit if you like! Steel, I may
> as well confess to you that a D.Sc. is my
> present aim & object & with other
> published work I think it ought not now to
> be far away — if only I can avoid
> having to pass the Hons B.Sc.

Extract from a letter to Bob Lawson, dated 14th September, 1911.
Holmes never did take a degree in geology.

of the Earth from an assumed temperature for the Earth, and instead he would calculate the initial heat of the interior from an assumed age of the Earth, which he considered to be fairly well constrained at around 1500 million years. He was clearly still concerned by Kelvin's arguments that the Earth had not been cool enough for life to have existed at the surface more than 20 million years ago, even if the Earth itself was considerably older than that.

While writing of this to Bob he suffered another bout of fever and was too ill to do anything other than think and write a few lines in his diary: *'Fooled on speculating on origin of Earth. Have concluded that the inner sphere of high density* [the Earth's core] *is of different origin to the siliceous acid and uranium-bearing exterior of much less density.'* Indeed, as we shall see, this remarkable idea is now exactly how some scientists do explain the Earth's core.

When the fever passed he continued the letter:

I shall now be able to write my article on the Age of the

Earth's Crust before Xmas and get it in the February number of Science Progress with luck. I intend writing all over the world to surveys and societies for material of known geol. age to analyse for U and Pb [uranium and lead]. *I am in hopes of gradually building up a geol. time scale and hope it might do for a DSc!!! There's conceit if you like! Still, I may as well confess to you that a DSc* [Doctorate of science] *is my present aim and object, and with other published work I think it ought not now to be far away.* He closed the letter to Bob: '*Well goodbye old man for tonight. I'm greatly looking forward to seeing you and my folks again and shall have a most gorgeous time.*

This is the last letter he wrote from Mozambique since others would be unlikely to arrive in England before he did. From here on we rely on the diary to tell us the rest of the story.

Barton and Holmes proceeded to Starey's camp to impart the news of the expedition's return to England, and here everyone stayed for their final two weeks. During that time Holmes' bouts of fever became more frequent and more severe, on one occasion his temperature rising to 105 °F: '*I shall be awfully glad to get on the boat again and safely out of this underhand sort of country*'. When not sick with fever Holmes spent the time working on maps and reports for Memba Minerals, and helping Mr Barton settle up the company accounts, which had got very behind.

He [Mr Barton] *really should have a secretary as the amount of clerical work in an expedition of this kind is enormous. There are four parties and two main establishments – Mosuril and Sawa – and all have rather complicated finance accounts for not only is money the currency, but also cloth of three kinds, wire, beads etc. All reports, bills, accounts, letters and documents of every kind must be copied and sent to the London office, so you will see that a good deal of work is entailed.*

Finally, after packing everything up, they arrived back at

Mosuril on the 9th of October and delighted in *the joy of a cool beer*. All they had to do was wait for the boat to arrive on their departure date of the 14th, then only five days away. But on the 11th Holmes was taken seriously ill again and this time the blood in his urine diagnosed blackwater fever. Unconscious, he was rushed by machila to the harbour at Mosuril, by native boat across the bay, and by rickshaw to the hospital in Mozambique. On his arrival the nuns who ran it had little hope for his survival. A report of his death was telegraphed to London.

Never believe what you read in the newspapers. With careful nursing Holmes gradually recovered, and eleven days later was discharged into the care of the Heseltons (the fine Yorkshire baritone from the Eastern Telegraph Company and his wife), who were packing up their house ready to leave Mozambique. He later recalled that it was being able to play the piano at their home, even though he only had access to it for a couple of days, that helped him to recover his inner strength.

Friday 20th October: Ordered a rickshaw to come and was allowed out [of the hospital] *for the first time. Still felt very weak in dressing, but enjoyed the ride to Heselton's immensely. Played piano tolerably well which gave me great delight. Saturday 21st October: Spent the day again at Heselton's and today being stronger managed the piano quite in my old style. Sunday 22nd October: Piano packed up today so no more music.*

During his illness Holmes had of course missed the boat home, so as soon as he was well enough he sent a telegram to Barton at Mosuril insisting he be allowed to leave Mozambique by the first available boat. But Barton never received the telegram. Unknown to Holmes, the day after he was discharged from hospital Barton was brought in, senseless with the gravest of malarial complications. But despite this, the next boat was due to depart on the 24th October and Holmes was absolutely determined to be on it. When he discovered that Barton was so dangerously ill he realised that he would be unable to access money owed to him that was locked in Barton's 'box'. He telegraphed Tasker, who was still in Mosuril, and asked him to send the money, but Tasker refused to open the box without Barton's permission. Holmes needed the money to pay the difference between the two boat fares and the ticketing office would not issue him with a ticket on trust in case Memba Minerals refused to pay the difference.

A couple of days later, with the *Palamaotta* now docked in the harbour, Barton was over the worst and Holmes was allowed to visit him: '*Saw Barton this morning – he wondered very much how he got there! He gave me permission to have his box opened for my money. Telegraphed this to Tasker but he still refuses to open the box. Difficulties in my getting away from Mozambique seem to bristle at every point.*' In the end he appealed to the British Consul, Mr Tothill, who suggested that he write a declaration making himself responsible for the difference in the fares, should the company directors demand it. This he presented to the ticket office who finally issued him with a ticket and allowed him to board. At the last moment before the ship left the eccentric Tasker arrived and thrust twenty volumes of Dickens into his hand for the journey. Holmes does not record whether he was grateful for this or not. In fact the only event he does record during the whole trip home was a severe bout of constipation which, on advice from the doctor, was embarrassingly relieved by a large dose of castor oil!

Having been away exactly nine months, he finally arrived at Southampton: '*Had early breakfast and telegraphed home. Left the docks at 9. Had lunch in London and did shopping. Arrived home at 8 p.m. Mother and all Lawsons at station. After arranging about luggage went round by shop to see father and had supper afterwards at Lawsons.*'

So Holmes was safely back home in the loving arms of his

family and friends, and no doubt he had the 'gorgeous time' anticipated. Although he lived for many years with recurring bouts of malaria, indirectly it may have saved his life as he was excused having to fight in the First World War. Even today malaria kills more than two million people a year, and is the greatest killer the world has ever known, so in 1911 Holmes was very lucky to survive.

The cost of the 1911 expedition was £15,995 13s 1d, adding to the financial burden already incurred by Memba Minerals, who ceased trading in 1913. The detailed accounts Holmes kept during the trip show that he made a profit on the venture of £89 7s 3d, almost one and a half times his scholarship allowance for a whole year, so in that respect he achieved his objective. But the most important outcome of the trip was that Holmes recognised that in order to understand the evolution of the Earth it was necessary to organize the rocks in a coherent and chronological fashion – and that the only way to do that was to date them. At the youthful age of twenty-one Holmes dreamt of developing a geological time scale and transforming geology forever; a vision that was to dominate his life for the next fifty years.

A Brimful of Promise

Here we are in the shadows of speculation and must await
the illumination of further discoveries.

Arthur Holmes

Margaret Howe was nearly thirty when she married Arthur
Holmes at Gateshead United Methodist Church on Tuesday the
14th of July, 1914. The daughter of one of the famous Howe
Brothers, a printing firm started very modestly by her grand-
father in 1863 which had grown to be one of the biggest employ-
ers in Gateshead, Maggie was the youngest of three children.
Her father, a master printer, was now comfortably retired but
Maggie still lived with her parents in their fine Victorian house
in Saltwell View, Gateshead, overlooking the park. Her brother
and sister had both married and moved away so it fell to Maggie
to stay at home and look after her parents.

It is unclear how the relationship arose, Arthur being five
years Maggie's junior, but he wrote to her from Mozambique so
they must have known each other for some while previously,
although at that time she certainly did not figure highly in his
affections. It seems likely that their parents were family friends.
Maggie did not have a university education and was certainly not
the geological companion of his dreams that he hoped Edie
might have become after he had advised her to study geology,
but she was a dark, pretty and vivacious woman with an

independent allowance. Having been 'squashed' by Edie maybe Arthur's attentions turned to Maggie during his visits home.

Most of the suitable young men of Maggie's age would have been married with a small family by the time they were thirty, so perhaps as she approached that age herself she was afraid of being left on the shelf. Arthur must have seemed a very glamorous figure after his return from Mozambique with stories of the African bush and how he had nearly died, so maybe it was then they fell in love. After the wedding the couple went to live in Chelsea so Arthur could continue his research, and exactly three weeks later, on Tuesday 4th August, war was declared.

In the early hours of the following day 600 Grenadier Guards took possession of the geology department at Imperial College (the Royal College of Science had been renamed Imperial College in 1910) and were billeted in all the rooms and laboratories. When Holmes arrived in the morning he had to step over sleeping soldiers lying in the corridors or propped up on the floor alongside cabinets and cases. Others were eating their morning rations using the tops of museum cases as tables. These cases were fitted with drawers filled with important teaching and research collections of minerals and fossils, all carefully labelled. Irreverently the soldiers deposited the remains of their rations into these draws and over time scraps of bread, meat and cheese gradually accumulated. Inevitably this attracted a large population of mice who indiscriminately ate the labels from the specimens along with the scraps. Much information was lost and the value of the collection as a teaching aid was enormously reduced, but it was a minor problem in the scale of this terrible war.

Almost at once the Imperial College authorities issued a notice that students would be granted leave of absence for the period of the war. They would be permitted to resume their studies at the point of interruption, the remaining portion of their fees being credited to them when they re-attended. Members of staff were also given leave of absence, their posts to be kept open for them, and any difference between their military pay and their college salary would be guaranteed to them by the Governors. Junior members of staff, however, were unlikely to benefit from this since their regimental pay, insufficient as it was, was still likely to exceed their pitiful college salary. College staff not eligible for active service were immediately deployed on domestic 'war work'. Holmes, still regularly suffering from debilitating bouts of malaria, was graded C3 (the healthiest grade being A1) and consequently deemed unfit for the armed services.

In the early months of the war, at the urgent request of the War Office, Holmes and several of his colleagues spent much time tediously making a one millionth scale map of Europe for Naval Intelligence. This involved the drawing of contoured maps of various parts of Italy, Russia, Finland and Scandinavia, depicting roads, railways and topographic features, and then converting the maps to a relief model of part of the western front, including Gallipoli. Later work, however, involved utilising his more specialised knowledge of geology. Before the war Britain had relied almost exclusively on German deposits of potash, a vital ingredient in fertiliser, so as the war progressed, acute shortages of potash were incurred due to the sudden cessation of supply from Germany. Working for the 'Sub-Department of Potash Production', Holmes played a significant role in researching and identifying alternative sources, directly contributing to the war effort by enabling more food to be grown.

For those on active service however, things were very different as was frequently recorded in The Phoenix, the college magazine which continued through the early years of the 'Great' war until, by 1917, there were no students left to write in it or for it:

The Great War has produced a profound change upon the life of the College. We have given of our best. Professors, lecturers, demonstrators, students and the working staff alike have responded to the call; and the numbers will continue to increase. There has been no hesitation. 'Our country first – and then back to College' has been the motto. Those of our *readers who have been unable to shoulder the gun are very proud – and envious – of those who have gone. We have suffered losses, but the sacrifice has been in a righteous cause.*

Such was the rhetoric of the day, although J. Forgan-Potts, who had just returned from France, was not so patriotic:

Wars, ever since time was, when compared, are much the same. One group of politicians annoys another group; then each convinces the populace that a war is completely justified: bells are rung; there is much show of cheap and ostentatious patriotism; then troops depart and khaki becomes the fashion. Presently many comparatively harmless people get killed – more people get killed on one side than the other.

By the autumn of 1914 a large number of Imperial students had already enlisted. As the War Office decreed that each man should enlist at the depot nearest to his home, many found themselves in the same regiment as their friends, as most students considered that 'home' was London even if their parental abode was elsewhere. As the war progressed young, homesick and bewildered students wrote to *The Phoenix* of their experiences. Royal Engineer G. S. M. Taylor was one:

I have been fairly in the thick of it since I came out here – in the trenches every night and also every other day; so I never seem to be out of them. It did not take me long to get used to the bullets. Sometimes they come unpleasantly close, and one touched my coat the other night. Both of the men I was with thought I was a 'goner', but I am a lucky sort of a chap.

My baptism of shell fire came the first whole day I was in a trench. We had 160 high-explosive howitzer shells right at our trench in three hours. I don't think anyone expected to come out alive, and a good many would have sooner died than have to sit and wait. Sitting huddled up in the trench, as low down as possible, you hear the report of the gun (about four to five miles away) and immediately hear the shell screaming towards you. The noise gets louder and louder; you don't know where it is until you hear the bang and roar, and then earth, sandbags or whatever it hits, go shooting up in the air. When this job was all over, some of our chaps were nearly mad.

I had quite an exciting experience this afternoon. I was coming down from a trench in daylight, and a sniper spotted me. Luckily, his first shot missed, and I flopped down in the communication trench. He had fifteen more shots at me, but could not quite get me, though he was doing marvellous shooting.

We had a bit of a 'do' the other day. The Germans broke into our gallery, but we hurriedly put in our charges, tamped and fired, and blew them sky high. Gee! it was jumpy work, but we managed it all right and my two men have both got medals. I see Ainslie is wounded. I was going to see him as I heard he was at 'Wipers'. I went into that place the other day to have a look at it. The Germans were shelling it with 14-inch guns, so I did not stay very long. They had just fired one into a house and killed 40 men. I have never seen such a sight.

As a consequence of the Imperial College students enlisting at the same depot, almost of all them were lost in the retreat from Mons. Taylor's luck finally ran out and he too was one of the millions who did not return home.

By 1916 Imperial College felt the full force of war-time conditions, and work was seriously impeded when all the resources of the Department of Geology and its staff were placed at the disposal of the Government. But prior to that, although conditions were not ideal, the fact that there was only a handful of students left to teach meant that more time could be devoted to research, and it was a particularly productive time for Holmes and his work on the age of the Earth.

While Holmes had been away in Mozambique, his friend Bob Lawson had obtained a brilliant result in his physics degree at Armstrong College, Newcastle. He was offered the post of 'prize' demonstrator as the reward for his success, which meant he could stay on at the university to continue his research whilst assisting a lecturer to 'demonstrate' to the students. In fact the job frequently entailed taking over from the lecturers while they got on with their research. When Holmes heard of this appointment he wrote a note that was both congratulatory and consoling to Bob: *'Allow me to cheer three times three. You have done just as well as you deserve. Your salary is not princely, but living at home and with half time for research it's not at all bad – and I should say you'll find yourself quite independent, though only in ordinary economics of course.'*

Back in London it had been much the same for Holmes. At Imperial College he too was offered a demonstrator post on a salary of £100 year, and despite having told Bob while in Mozambique *'I'm afraid this life is spoiling me for anything again like my old struggle against poverty. I feel much more on the £1,000 a year scale of living'*, he nevertheless settled down *'to normal ideas again after a week or two at home.'*

Immediately on his return from Mozambique, Holmes had decided to write a small book on *The Age of the Earth*. With frustration mounting at the entrenched attitudes of established geologists, he wanted to explain to them, and tell the world at large, about radioactivity and his ideas for a geological time scale. At the same time he admitted that confidence in pioneer work in radioactive dating *'has been shaken by the advocates of the geological methods of attack. The surprises which radioactivity had in store for us have not always been received as hospitably as they deserved. With the advent of radium geologists were put under a great obligation, for the old controversy* [with Kelvin] *was settled overwhelmingly in their favour. But the pendulum has swung too far, and many geologists feel it impossible to accept what they consider the excessive periods of time which seem to be inferred.'*

With his enthusiasm and convictions tempered by a fear that there might still be something in the hour-glass methods, he took pains to review all the facts for the benefit of his readers, but still concluded that *'the uncertainties* [of the hour-glass methods] *are too many and too great'*. Having then reviewed all the evidence in favour of radioactive methods, he warmed to his theme: *'As yet it is a meagre record, but, nevertheless, a*

record brimful of promise. Radioactive minerals, for the geol-
ogist, are clocks wound up at the time of their origin [and] . .
. we are now confident that the means of reading these time-
keepers is in our possession. With the acceptance of a reli-
able time-scale, geology will have gained an invaluable key
to further discovery. In every branch of the science its mission
will be to unify and correlate, and with its help a fresh light
will be thrown on the more fascinating problems of the Earth
and its Past.' We can almost see him marching forward on his
crusade, banner held high!

With his book on *The Age of the Earth* just completed, an event
occurred that, indirectly, was to have a great influence on his
work: Bob Lawson was offered a post at the recently opened
Radium Institute in Vienna. Established in 1910 with a contribution from an anonymous donor – about a million pounds in
today's money – the Radium Institute in Vienna was built to rival
work being done on radioactivity in France and Britain. The
benefactor reasoned that since Austria housed the important
Joachimsthal uranium mine, from where Marie Curie had
obtained her uranium and radium, and since there was such
world-wide interest in radioactivity, then Vienna should have a
Radium Institute of its own to compete with the leaders in this
science. It immediately attracted some of the best brains in
Europe, and in 1913 Lawson found himself working alongside
people such as Viktor Hess and Georg Von Hevesy, both of
whom went on to win Nobel Prizes, and Fritz Paneth, who was
later to play his own role in the Dating Game but now, like
Lawson, was just starting his career. It is an indication of
Lawson's ability that he was invited to work at the Institute and
he soon became a valued member of the international team, but
with the outbreak of war he found himself unable to get home
and was thus detained in Vienna for the next four years. As
Austrian nationals were called up and student numbers in the
Institute declined, Lawson, with the help of some Polish
refugees, became a crucial figure in holding the department

together and continuing to teach the few students who remained. This was no bad thing since, like Holmes in the early war years, he too found extra time for research.

Holmes had realised, just before he left for Mozambique, that one of the potential problems with the uranium–lead method of dating minerals was in precisely evaluating the amount of lead that resulted directly from the decay of uranium. If some of the lead he measured in a sample was what he called 'ordinary' lead (lead that had been around since formation of the Earth and so was already present in the mineral when the lead from uranium started to accumulate) then the age deduced from such a sample would appear to be much older than it actually was. There were still many geologists who opposed the idea of an ancient planet and who considered that all radioactive techniques were totally unreliable, citing with pleasure the problems that had been encountered with helium, so they would be only too delighted to find a similar flaw in the lead method. But how to tell these two types of lead apart? Just before Lawson left for Vienna the two friends had been discussing this very problem in the light of the most recent 'wild miracle'.

By 1913 the nucleus of the atom had been discovered (by Rutherford in 1911) and so it was known that the atom consisted of two strongly contrasting portions – a nucleus, surrounded by an intense electric field of electrons – but it was still considered that elements could be distinguished, one from another, by their different atomic weights. Thus two atoms with different weights were considered to be two different elements. So when Frederick Soddy, now working in Glasgow on the decay products of thorium, found that it was impossible to chemically separate one of thorium's decay products from thorium itself, although the two had different atomic weights, he was considerably perplexed. And the more he examined the problem the more he found other pairs of elements that were chemically indistinguishable from each other, although they too had different atomic weights.

It took a while for Soddy to realise that the reason why these pairs of elements were inseparable was because chemically they were the *same* element, despite having different atomic weights, since each pair had the same number of orbiting electrons around the nucleus, determining their chemical identity. In other words, what appeared to be two separate elements actually turned out to be two varieties of *one* element. Soddy termed the different varieties of the same element 'isotopes' (from the Greek *isos*, same, and *topos*, place) because, as he now recognised, all varieties of any one element would occupy the same place in the periodic table – a classification which summarises the major properties of the elements and enables predictions to be made about their behaviour.

Today we know that the nucleus itself is comprised of two types of particle – protons and neutrons – and that an element's position in the periodic table is dictated not by its atomic weight but by its atomic number, which is the number of protons in the nucleus (and the number of electrons orbiting in the surrounding field). It is the number of neutrons in the nucleus that varies from one isotope to another, and it is this that determines an isotope's atomic weight. The neutrons and protons added together give the isotope number.

For example, an atom of lead derived from the decay of uranium (238) is called 'lead 206' because it contains 82 protons (and 82 electrons) and 124 neutrons (82+124 = 206), hence '206' is the isotope number. Similarly, the isotope derived from the decay of thorium (232) also has 82 protons but 126 neutrons, and so it is called 'lead 208'. Since all the isotopes of lead have 82 orbiting electrons they cannot be distinguished from each other chemically, but because each has a different number of neutrons, their atomic weight is different and so they can be separated physically.

The concept of isotopes as put forward by Soddy was yet another radical departure from accepted understanding about the atom and it was some while before he was brave enough to put his ideas into print. But no other discovery has subsequently proved so critical to radiometric dating, and the significance of it was immediately recognised by Arthur Holmes.

So far he had assumed that any 'ordinary' lead in a rock was present in such small quantities as to be insignificant when compared with the amount produced by the decay of uranium, but in fact it was impossible to tell. To muddy the waters even further, Soddy had shown that an isotope of lead was the final decay product of thorium as well, so now there were three possible sources of lead – uranium, thorium and 'ordinary'. It really was a problem that needed to be resolved, and clearly isotopes were the path to enlightenment. Unfortunately, at that time, the only way to distinguish one isotope of lead from another was by measuring their atomic weights, painstaking work that required even more fastidious care than separating uranium and lead.

But the rewards for success in this field were high, and by 1914 the team working in the Vienna Radium Institute had become world experts on atomic weight determinations. So despite the difficulty of communications with Austria during the war – there frequently weren't any – at the instigation of Holmes, Lawson and the Viennese team accurately determined the atomic weight of the three lead isotopes in a number of samples and showed that the results were both consistent within themselves, and with other workers in the field. So now when a new mineral was dated it became possible to calculate the proportions of each type of lead from the atomic weight of the sample and adjust the age accordingly, ensuring that it was not over-estimated. It was a significant breakthrough and Holmes was gratified to report: *'that the uranium–lead ratios may be used as before and with greater certainty for the determination of geological time and the gradual construction of a complete geological time-scale'.* It must have been quite a relief to find that after all their pervious hard work, the system still worked with only minor adjustments required. Or so they thought.

Unknown to them at the time was the fact that uranium itself also has three isotopes, and although more than 99% is uranium 238, uranium 235 is also significant because it decays much faster than 238. And yes, the end product of uranium 235 is yet another stable isotope of lead! So, when the Viennese team were measuring what they thought was 'ordinary' lead, it turns out that what they were actually measuring was the lead produced from the decay of uranium 235, which had not yet been 'discovered'. Thus when Holmes corrected his uranium–lead ages for the presence of thorium-derived lead and what he thought was 'ordinary' lead, in fact the 'ordinary' lead was still unaccounted for.

And so science progresses. Three steps forward, one step back, but each little bit adds to the bigger picture. With this new understanding of lead and its isotopes, and the ability to adjust previously determined ages to apparently account for the presence of 'ordinary' lead, the oldest rocks from Mozambique were now judged to be 1500 million years old. And because the Earth had to be older than the oldest rock it contained, the age of the Earth was deemed to be at least 1600 million years. But not everyone was convinced.

In March 1915 Henry Shelton, school teacher and journalist on 'scientific and philosophical subjects', gave a talk to the Geological Society which examined 'The Radioactive Methods of Determining Geological Time'. Shelton was clearly not a convert and as a consequence, hardly an authority on the subject. Phrases such as

The attempt to assess exact, or even approximate times by means of lead ratios is premature and entirely invalid' and 'radioactive experiments . . . are not sufficiently important in themselves to be authoritative against the balance of the evidence derived from other lines of investigation

were bound to antagonise Holmes. Although generally a quiet, reserved and rather shy individual, Holmes never had any compunction when it came to standing up and defending his cause. With his deep knowledge of the subject he was driven by a conviction that no matter what ever anyone else said, the ages he had determined were at least within the right order of magni-tude. The Earth was not less than a hundred million years old as suggested by Shelton, but at least a thousand million years old.

Unhappily, all radioactive procedures were suffering from the damage inflicted on them by the helium method, which Holmes had shown gave ages almost half those measured by the uranium method, so it was against this background that he stood up to defend his case for an ancient planet. Always polite, he at first welcomed the opportunity to discuss the matter, *'But find I cannot agree with Professor Shelton's generalisation that the radioactive methods were not authoritative in their own right. If that were the case, it would surely be necessary to give up any serious attempt to construct a geological time-scale'.* It obviously never entered his head that other people might think it was not necessary to construct a time scale! Advocating a positive approach to the helium set-back, he even turned it round to have a dig at his opponents:

We must not moan over the apparent diffculties with which the geologist has been faced since the advent of radium. If at present some of our ideas are mutually incompatible, the discrepancies do not demand a wholesale rejection of the facts, but simply a re-interpretation of the fundamental hypotheses on which so many of our doctrines seem to hang.

But his reasoning fell on deaf ears and his opponents continued to dismiss all radioactive results as simply too unreliable, this time citing inconsistencies within the uranium–lead method itself. Feeling more and more backed into a corner Holmes replied:

When inconsistent ratios are obtained there is either evidence of extreme alteration of the minerals or it has been contaminated due to the presence of 'ordinary' lead. However, with the new knowledge of isotopes from atomic weight determinations, corrections to anomalous results can now be made.

Shelton was dismissive:

I consider the problem lies in the unreliability of determining minute amounts of lead and in my opinion the suggestion that lead formed by the degradation of uranium has a different atomic weight from 'ordinary' lead is contrary to the trend of chemical investigation, and should not be regarded as established.

Fifty years later Holmes was to recall the occasion:

I was being violently attacked by the reader of a paper who insisted that the Age of the Earth must be less than 100 million years old. In the discussion that followed I had occasion to refer to the isotopes of lead, then newly discovered. But isotopes did not seem to have been heard of in that audience. The reader of the paper insisted that all atoms of lead must have the same atomic weight, and I found myself in an exasperated minority of one.

This was to be a common occurrence for many years to come.

A few months later the annual meeting of the British Association was held in Manchester, despite the war, and the inevitable discussion about the age of the Earth ensued. Chaired by Rutherford and with Soddy and Joly in the audience, Holmes again presented his arguments for an Earth at least 1600 million years old. Joly, still strongly in favour of his age of the Earth estimated from the sodium in rivers, recognised the value of radiometric ages but rejected the uranium–lead method, accepting only the younger dates provided by the helium method. He argued fiercely in favour of these because they were more in keeping with his age of the oceans. While Holmes was used to these arguments and now fully expected all geologists to urge caution at the idea of the Earth being so ancient, he was not expecting Soddy of all people to take up a similar position:

It is hoped that geologists would not be in any immediate hurry to decide between the geological and radioactive estimates of the age of the Earth. Owing to the element of uncertainty about the initial stages of disintegration . . . there might well be unknown factors still to be discovered sufficiently important to bring the two methods into closer agreement.

Wearily Holmes reflected that there was still a very long way to go.

Liquid Gold in Yenangyaung

There have always been optimistic operators willing to risk
their capital and take a chance by sinking 'wildcat' wells on
sites selected for some quite unscientific reason.

Arthur Holmes

By the end of the war Holmes was still at Imperial College, still
only a demonstrator and still on a salary of £150 a year, despite
having published three books and gained a significant reputation for his work on radiometric dating. His finances were
permanently under pressure such that when Maggie gave birth
to their son Norman within two weeks of Armistice Day in 1918,
they became critical. He had tried to get other jobs but despite
being awarded the doctorate dreamt about in his letters from
Mozambique, somehow no job had materialised either before,
during or after the war. Watts had tried to get him a position as
lecturer in petrology at Oxford *'but naturally failed!'*; he had
testimonials from many eminent geologists for his application
to Cardiff; but in 1919 he could not even get a teaching job at
Aberystwyth. He wrote to Dr Prior:

*I am glad to tell you that Aberystwyth failed to appreciate
my qualifications for the geology post and appointed a student (Welsh!) instead. The department is very small and
crowded and its chief objective appears to be to train girls to
pass examinations to be teachers. So I am well out of it!*

He was, nevertheless, still broke.

To make matters worse, at the end of the war Bob Lawson had left Vienna for a lectureship in physics at Sheffield University, so Holmes' access to the Viennese experts on atomic weight determinations was cut off. This seriously hampered his age dating work. Already the necessity for complex chemical separations of radiogenic lead imposed severe limitations on the uranium–lead method and restricted its application to rare minerals containing appreciable amounts of uranium. The problem was compounded by the presence of both uranium and thorium derived lead and also by the existence of appreciable amounts of 'ordinary' lead. No longer was it acceptable simply to measure the total amount of lead in a sample, it was now necessary to know which isotopes you were dealing with and this was only possible by determining their atomic weights. So even with access to the Viennese lab, the delicate and extremely accurate atomic weight determinations only complicated further the already complex chemical analyses. It was just too time-consuming for everyone concerned and after the war when the students returned and time for research was severely curtailed, the method largely fell into disuse. For a while Holmes' dream of building a geological time scale became less tangible as more urgent matters required his attention. By 1920 his financial situation had become so acute that it was imperative he get a better paid job.

In 1911 a pioneer series of lectures on oilfield geology had been given at Imperial College, and within a very short time an Oil Technology division of the Geology Department was borne. Many of the students from this course found jobs in the booming oil fields of Burma and on their return would tell stories of their adventures. There seemed to be no alternative: late in 1920 Holmes resigned from Imperial College and accepted a job as Chief Geologist to the newly formed Yomah Oil Company (1920) Limited.

The original Yomah Oil Company had been initiated in 1913 when its three British directors bought 500 acres of land in Burma's Minbu Oil Field from a Mr Hugh Porteous Cameron of Rangoon, for the lavish sum of £124,693. The vendor was to receive £25,000 in cash and the rest in shares, but £20,000 of the cash would be retained until the property was producing four hundred barrels of oil a day. Mr Hugh Porteous Cameron of Rangoon must have had a lot of faith in his land! To be fair, the Minbu District did look promising. Production of oil had increased from 18 000 gallons a year in 1910 to just under four million gallons by 1912, and 4000 barrels a year was being produced from a well right against Cameron's boundary. However, Cameron had already drilled six wells on the property, furnished the site with the latest equipment, laid a pipeline to the Irrawaddi river five miles away and purchased land there for oil storage, all without any significant amounts of oil actually being produced. Before production could commence he had run out of funds and sold out to the Yomah Oil Company who retained his services as General Manager. Three years later The Yomah Oil Company was in the hands of the receiver.

But in March 1920 The Yomah Oil Company (1920) Limited was resurrected from the ashes of its dead parent, bought at an advantageous price by the infamous Lim Chin Tsong, who was known throughout the Burmese Oil Industry simply as LCT. Until 1919 the charming, inscrutable and entrepreneurial LCT had been, amongst his many other roles, an agent for the Burmah Oil Company, the oldest of all British oil enterprises. From an office in China Street, Rangoon, with its celebrated telegraphic address of 'Chippychop', LCT ran a network of oil distributors that extended over the whole of Burma, which was then part of India and a British colony. Throughout Burma the Chinese had a virtual monopoly of retailing through the general stores they ran in every town. These were supplied by LCT with kerosene for domestic use, which they in turn sold on to the local villagers who arrived with their own containers, usually beer bottles, for filling at the local shop.

LCT was supposed to pay Burmah Oil, on a monthly basis, all the money he received for sales to these outlets, after taking a small commission for himself. But shortly after acquiring sole agency rights to sell Burmah Oil's products throughout the country, LCT fell behind with his payments. Over several years the debt accumulated and when the amount owed ran to several hundred thousand pounds, a sizeable sum now let alone in 1912, the company chairman himself came out from Glasgow to assess the situation. Although LCT spoke excellent English he could neither write nor read it and kept all his accounts in Chinese, much to the frustration of the Burmah Oil's dour Scottish accountants who tried to decipher what he owed them. Strict instructions were issued that he was not to exceed the limits placed on the debts he could run up. It had little effect. Somehow he managed to keep the accountants at bay for another seven years with promises of payment, some of which materialised, most of which did not.

Outwardly LCT was an extremely wealthy man. He owned steam ships, a rubber plantation, a match factory, a tin mine and a coal field, and had built a huge palace in the shape of a star on Kokine Hill overlooking Rangoon. The pagoda-style central tower rose graciously above the main reception hall. Rooms off this formed the points of the star. It was luxuriously furnished with oriental treasures, and he and his wife entertained distinguished visitors to tea and dainty sandwiches of seaweed jelly, considered a particular delicacy. Accordingly he moved in the highest echelons of Burma's business and social life, where he had very good connections. Through these connections he had helped secure the rights for Burmah Oil to drill for oil in Upper Burma, without which the company's fortunes may well have been very different. Because of this there was a tendency for Burmah Oil to be lenient with him.

LCT's inability to pay Burmah Oil was because he was in

Lim Chin Tsong's Palace on Kokine Hill in Rangoon.

financial difficulties with his other enterprises and was using the money owed to Burmah Oil to stave off his creditors. As his business fortunes rose and fell with erratic irregularity, he was alternately millionaire and pauper. He was also an inveterate gambler. But by 1919 Burmah Oil could take it no more and told LCT that they were cancelling his agency agreement. They courteously offered him the option to resign first, which he did, still promising to repay all the money he owed. Later that year, despite, or perhaps because of, his financial crisis LCT travelled to England for the first time, a trip he had always promised himself. He stayed for three months spending £30,000 at Harrods on furnishings for his palace, and amusing himself in the West End. To pay for these extravagances he took from Burma a boatload of Burmese commodities to sell in London, the profits from which were also to repay his debts. But inevitably the money became side-tracked.

Within the small circle of foreigners in Burma, everyone knew everyone else, and they all knew LCT. For many years Hugh Porteous Cameron had been Chief Engineer to Burmah Oil so he and LCT had known each for some time. At some point during his business activities in Rangoon, LCT renewed this contact with Cameron who was keen to recover some of his losses from the Yomah Oil Company. LCT did not so far own an oil company and was attracted by the idea of such an enterprise, while Cameron was presumably not party to the real reasons why LCT had parted company with Burmah Oil, or the fact that he owed them such large amounts of money. Cameron undoubtedly saw LCT as a financial saviour. Consequently, on LCT's return to Burma from England, they re-launched Yomah Oil Company (1920) Limited with the revenues derived from LCT's dealings in London. Cameron was back in place as General Manager and LCT's sons, who had been educated in England, were named as directors.

This development was noted with concern at the Burmah Oil Company who obviously viewed each new entrant into the oil arena as a competitor, and who saw hopes of ever recovering their money from LCT rapidly receding. Nevertheless, optimistic instructions were issued from Glasgow to the new agent in Rangoon, 'we quite agree with you that so far as possible it is very desirable that you should endeavour to maintain friendly relations with Mr. Chin Tsong provided this can be done without preventing you from recovering promptly from that gentleman the full amount of his indebtedness to the Company!'. LCT never did pay back Burmah Oil.

In January 1920, Arthur Holmes, who had been impressed with the grand lifestyle exhibited by LCT when they had met at LCT's hotel during his visit to London, knew little of Yomah Oil's rather unfortunate history. He gladly signed his contract that promised him a salary six times greater than he was currently getting, even though that had just been increased to £200 a year, and looked forward to finally being on the scale of a 'thousand pounds a year man' that he had dreamt of back in Mozambique. In addition to this salary he was promised a first class ticket on the boat out, all expenses paid, and a house when he got there so that his wife and young son could go with him. At the end of his three year contract, should both parties wish to renew it, he would be entitled to six months holiday 'out of India'. He was confident that the company was going to treat him well. Arthur, Maggie and their son Norman left for Burma in November 1920.

The Irrawaddi River in Burma is one of the longest rivers in the world and is navigable for over a thousand miles. With its source high up in the snows of the Himalayas, during the melt period more water flows past Yenangyaung in twenty four hours than passes under London Bridge in a whole year. Yenangyaung was the capital of the oil fields and where Arthur and Maggie were to live. Although the Minbu field was on the opposite bank of the Irrawaddi, some 40 miles to the south west of Yenangyaung, the two places were easily linked by the river. However, facilities for women at Minbu were distinctly limited and unattractive, and since the expatriate clubs, and hence the social life, were all in Yenangyaung few women chose to live in the field. Situated two hundred and fifty miles north of Rangoon, just over half way between Rangoon and Mandalay as the river flowed, the only way to reach Yenangyaung in those days was up the Irrawaddi by paddle steamer. Twice a week mail steamers sailed from Rangoon to Mandalay. The complete journey, with stops at all the important towns, took a week; Yenangyaung was four days.

The first class accommodation consisted of eight two-berth cabins with bathroom, arranged in rows either side of the foredeck with a dining salon in between. While this was quite commodious, the majority of travellers were 'deck' passengers who kept up a tremendous hubbub during the day with much to-ing and fro-ing as they came and went at each port of call, loading and unloading their cargoes. Between ports they squatted on the steel deck chatting, chewing betel nut and smoking white cheroots which they passed from one to another, and if they were going further afield than could be covered in a day they would unroll their plaited bamboo sleeping mats and sleep out on deck. Buddhist priests in their saffron robes wandered about beating their gongs and accepting a share of their compatriots frugal meals.

As the steamer sailed out of Rangoon the first view seen beyond the muddy brown banks of the river was a sea of emerald green rice fields shimmering to the horizon. Between the wars Burma was the world's largest producer of rice. Occasional villages could be spotted from a distance by the toddy palms, distinctive for their long trunk topped with a tuft of green, looking much like the pom-pom on a circus horse, that fringed each one. As they approached the villages they could see Burmese women with their children bathing in the river or washing the family's clothes, as water buffalo, used for many an agricultural task, wallowed in the shallows with only their nose and horns visible above the water.

Gradually, as they steamed further north, the landscape became more undulating as the jungle-covered foothills of the Arakan Mountains to the west were approached. Near Prome, the end of the railway line from Rangoon and where many passengers embarked for the oil fields, the river cut deeply into the hills creating high rocky cliffs festooned by the jungle that gracefully draped itself over the edge and into the water. Brilliant birds of every colour flashed amongst the dense green foliage, settling in their hundreds on the overhanging branches, and on every promontory a white pagoda perched in the hope of earning Nirvana points for its Buddhist builder. Beyond Prome the bamboo reached heights of 30 feet and the Pegu mountains to the east also came down to drink at the Irrawaddi. The two mountain ranges, now towering either side of the steam ship,

The Yenangyaung Oilfield.
On the back of this postcard Arthur writes that the photo was
taken several years earlier, and that the density of derricks was
now (1921) even greater.

shed their surplus water via streams and waterfalls, creating a
scene of grandeur and tranquillity. In this way Arthur, Maggie
and two-year-old Norman slowly made their journey to
Yenangyaung, absorbing something of this country that was to
be their home for the next few years.

Once beyond the jungle the transition to the parched scrub
of the dry zone was rapid, dramatically marked by the complete
lack of trees and green vegetation. The bamboo became dry
and brown, barely reaching more than 10 feet high, and the
landscape appeared barren in comparison to the jungle. As
Yenangyaung was approached the cliffs rose abruptly from the
east bank and a dramatic serrated horizon appeared, looking at
first like a forest. As they came closer it could be seen that the
forest was not of trees, but of two thousand wooden oil derricks
strung out along the three mile crest of the oilfield. Arthur and
Maggie disembarked somewhat apprehensively at the port of
Nyanghla, with the help of coolies and bullock carts to transport the baggage.

The beginnings of the oil industry in Burma are known to go back to at least the thirteenth century. By the eighteenth century it was the largest producer of oil in the world, and it was not until the middle of the nineteenth century that it was overtaken by the United States. At that time the rights to sink wells were shared by 24 families known collectively as the Twinzayos (literally 'well eaters' or those who lived from the proceeds of well ownership) and these rights were handed down through the generations from male to male, or female to female. In 1885 the last but one of the Burmese kings, King Mindon, acquired nearly two hundred wells, a large number of which came as a dowry, and it was these Royal wells that were first leased to the Burmah Oil Company when it was formed in 1886. So for twenty years, until 1906, the Burmah Oil Company and the Twinzayos shared the oil bearing areas of the Yenangyaung field, but by 1908 the Twinzayos realised the benefits of allowing outfits with modern equipment do all the work, and leased wells to five other companies. The race was on. In 1895 a well site had sold for 100 rupees, by 1908 it was worth 60 000 rupees, approximately £4000.

The first mechanically drilled well in Burma was completed in 1887. Until that time oil production had been obtained from hand dug wells above natural seeps of oil occurring at the surface. Shafts for the hand dug wells were started by digging pits about five feet square which were shored up by timbers. With increasing depth the well digger himself and, of course, the spoil to be removed, were hauled to the surface by 'coolies', mostly women, who ran down an inclined path dragging behind them a rope that passed over a beam spanning the well head and was attached to the digger. No one seems to have thought of a windlass. When the pit became so deep such that very little light could penetrate, a mirror would be angled so that the sun's rays were beamed to the bottom; obviously before electricity no other form of lighting could be used without a severe risk of fire.

Below two hundred feet the amount of gas in the well prevented the digger from remaining there for more than a minute at a time and between each descent he needed half an hour's rest to recover from the effects of the gas. In such conditions the digger managed about twenty descents a day; progress was very slow, taking nearly a year to reach three hundred feet, and the mortality rate from gas poisoning was very high. But with the introduction of four-gallon tin cans for transporting kerosene, a crude yet ingenious diver's helmet could be made from an inverted can placed over the digger's head, connected to the surface by a rubber hose through which air circulated from an air pump operated by a member of the digger's family. With this significant improvement the digger could stay down the pit for up to an hour and depths of four hundred feet could be reached. Nevertheless, by the turn of the century production from a hand dug pit was still only twenty gallons a year. Although they rapidly became overtaken by mechanical wells, many hand dug pits were still in operation at the time Holmes was working for Yomah Oil.

By 1920 oil production in Burma was at its peak and competition for this liquid gold became quite frenzied. Many of the smaller companies had prospered during the war, with the Government paying high prices for whatever oil could be produced, and they were now in a position to branch out and sink new wells. But the oil bearing province at Yenangyaung covered an area of only one and a half square miles, stretched along a narrow ridge just three miles long and half a mile wide. It was divided into two oil bearing 'reserves', the Twingon and Beme reserves, and space was running out. Historically the Twinzayos had divided up the reserves into circular sites, each of which was only sixty feet in diameter. As this system continued a congestion of highly inflammable wooden derricks arose, so close to each other that it was often unnecessary to touch the ground when walking from one to another.

This disorderly collection of wooden derricks, settlement tanks, pumps and steam boilers strewn amidst a labyrinth of gas, oil, water and steam pipelines gave the newcomer a sense of indescribable confusion, and finding one's way around could be a nightmare. Fortunately Upper Burma lies within the dry belt where rain was infrequent, but when it did fall landslides of the unconsolidated muds became a frequent cause of serious problems. Rain gouged deep, steep gulleys into the soft clay making the movement of personnel and bullock carts incredibly difficult. And oil was everywhere. It covered everything with a thin, slippery veneer, and its smell hung in the air all year round such that you stopped noticing it, except when you left.

Work as Chief Geologist to Yomah Oil was demanding. By the time Holmes arrived the company had already drilled twenty-three wells in the Minbu field, and was exploiting a rather poor coal field in the hope of finding something more extensive. They also leased several well sites on the Twingon reserve in the Yenangyaung field which were reasonably productive, generating between them up to a thousand barrels a day. Since Yomah Oil did not have its own storage or refining facilities at Yenangyaung, the oil was bought from them in 1920 by Burmah Oil at 8 rupees per gallon, generating a handsome turnover of almost £200 000 in that year. A considerable price increase to 11 rupees a gallon the following year marked a continued increase in demand that was outstripping supply, but in fact 1921 marked the beginning of the end for oil production in Burma.

In complete contrast to Yenangyaung the Minbu field was located on the edge of the jungle, which was home to a large variety of wild animals including tigers, leopards and elephants. Minbu also boasted spectacular mud volcanoes. Globules of viscous oil would rise up through liquid mud on a column of gas, bursting with some force as they reached the surface and splattering the vicinity with oil. But despite this apparent evidence of oil the Minbu field was far less productive than Yenangyaung, and well No. 24 was the great white hope on which everyone in the Yomah Oil Company pinned their dreams – as they had done on the previous twenty-three. As many of the wells were now penetrating to depths of up to three thousand feet Holmes was frequently required to visit well No. 24 to determine whether the drillers had found the oil bearing sands, or whether they should continue drilling, but his main task as Chief Geologist was to locate and evaluate the potential of new fields. That required a great deal of travel from one place to another on slow modes of transport such as the paddle steamer, bullock carts or pony which made for long days in the field and frequent frustrations.

While Arthur was away Maggie stayed behind in Yenangyaung, enjoying the facilities offered by the Yenangyaung Club, while two-year-old Norman was largely cared for by the servants. Office hours when the men were out of the house were 6 to 10 a.m. and 2 to 4 p.m., but beyond organising the daily meals a wife could not do any housework or shop in the bazaar without losing the esteem of her servants. Wives would often congregate in the morning at the Club for a game of bridge or tennis, or a round of golf under the somewhat unusual conditions – the fairway was bare gravel and the green was an oiled patch of sand known as a 'brown'! Gardening however, was allowed by the servants as long as the Memsahib did not actually do any of the work herself, and thousands of gallons of water a day, liberally poured over the borders, ensured a dramatic and continued pageant of colour, the wives competing with each other for the best display.

The only other social amenity was the Yenangyaung cinema, located in Yenangyaung town, where silent films were shown to the accompaniment of a piano. The expatriate communities would often organise dinner parties that went on to the cinema, taking with them their own liquid refreshments in an ice box, and a servant to pour it for them. When Arthur was home he and Maggie would frequently entertain their friends for dinner, or be invited out, and regular 'soirees' were held where Arthur played the piano or they listened to the gramophone, records becoming like gold dust. Informal dances were organised every Saturday evening at the Club and once a month a more formal event was held when fancy dress was often *de rigueur*. The Sunday breakfast following these occasions was a major social event where the evening's gossip was told and retold to those unfortunate enough not to have been there. It was a hard life.

In the first year Holmes travelled widely with his colleagues Stanley Hunter and Dudley Stamp, later to became a famous geographer, as far as the borders of India and Tibet, in the search for new oil and coal fields, making geological reconnaissance maps as they went. Known to the locals as 'Rock-pidgin-man', Holmes was elevated to the position of 'Number-one-top-piece-rock-pidgin-man' when, in 1921, he was promoted to 'Up Country' Manager and Geological Advisor. This increased his responsibilities to full charge of the oil and coal field staff, as well as the geological surveying and prospecting departments. However, with disappointing returns from well No. 24, it must have been obvious to all concerned that the company was sliding into financial difficulties. As a result, early in 1922, Holmes decided to close down the coal field. Enormous expenses lay ahead of them if they were to make the field viable. It would be necessary to extend the railway line for transporting the coal that was currently conveyed by bullock carts whose prices were rising astronomically, and build a new camp for the workers, to say nothing of the equipment that was needed, the rent to be paid and the salary bill of 5000 rupees a month. It was clear the company was not in a position to finance such an investment.

By March the drillers on the oil fields were threatening to close down the wells unless they were paid the several months wages owed to them. In an urgent telegram Holmes was summoned to Rangoon for meetings with Cameron and LCT. Five days later, the fastest he could get there, he was graciously received at LCT's palace. He was conducted from room to room to see the valuable draperies, luxurious rugs and bed covers, and the fabulous collection of silver, gold, jade and marble ornaments that Holmes particularly admired. Tea was served and small talk prevailed. Business waited until Cameron arrived next day.

It transpired that Cameron and LCT were attempting to raise capital in London by issuing a hundred thousand shares in the company at one pound each. This money, LCT assured them, would solve all the company's financial problems and allow them to get on with the business of finding oil. In his charming and seemingly candid manner LCT convinced Holmes that there was nothing to worry about and that in a short while they would all be rich men, so Holmes spent the rest of the week in the Chippychop office working with Cameron on preparing a report for the share issue prospectus. Satisfied that all would be well and with a promise from LCT that salaries would be paid in a couple of weeks, Holmes left Rangoon for Mandalay, where he met Maggie and Norman off the steamer, who joined him for a rare holiday. They toured the palace and sights of Mandalay and then visited the bazaar, where Arthur bought Maggie a beautiful green jade necklace, and a pair of chased bowls on cobra stands for Bob Lawson's wedding present.

That evening they left for Kalaw, a small town in the cool of the hills that was a short train ride from Mandalay and an escape from the heat and smell of Yenangyaung. Maggie suffered from asthma and found the dust and oppressive heat of the oil fields difficult to cope with, and Arthur had had several attacks of malaria which kept him in bed for days at a time as the fever raged through his body for two or three hours each day. They both needed a rest. Norman, however was a bright and lively three year old with chubby cheeks and golden hair that attracted the attention of Burmese and expatriates alike everywhere they went. Arthur had seen little of his son during their time in Burma but when he was home he greatly enjoyed taking the boy on walks, encouraging his interest in the rocks and wildlife around him from an early age. He was a loving father and often sent postcards home to friends and relatives signed with Norman's name, knowing how they would be missing him. He had wanted to provide the best for his son, so he must have found their current financial difficulties a great worry and disappointment after his early expectations of the riches to be made in the oil fields of Burma.

But the demands of the job did not allow him to stay in one place for long and by the end of the week he had left Maggie and Norman in Kalaw, where they were to stay for another month, and was back in Yenangyaung to reiterate to the oil field staff the promises made by LCT regarding salaries. But when the promised money was not forthcoming, Holmes went back again to Rangoon. This time he managed to obtain 1000 rupees for himself, equivalent to a couple of weeks salary, and 500 each for Hunter and Stamp to cover immediate needs, with promises of more to come, but the other salaries were still not paid. It was the last straw for Hunter. Following a severe bout of malaria, and with his young wife pregnant with their first child, he had had enough of promises. Hunter resigned from Yomah Oil as soon as he was well enough to put pen to paper.

Holmes carried on in a frenetic search for possible new oil finds. Endless days were spent travelling on trains and boats rushing up and down the country from one site to another. Everyone seemed to want his opinion on this geological structure, that possible coal seam, or this problem at the well site. Day after day he records being in the field looking at the geology, travelling huge distances over difficult terrain, most of it on foot or by pony. There was no time for thinking about a geological time scale, the age of the Earth or the problems of lead isotopes. Those days seemed very far away now.

Every time he was near a telegraph office he sent a telegram to LCT asking for money. At the beginning of June he received a reply to the effect that no money was available and he was forced to sell 200 of his own shares to meet his most urgent financial requirements as the bank would not extend his overdraft. Several weeks and numerous telegrams later another reply came saying that Cameron was 'too ill to attend to business' and LCT could not be contacted. Incensed by their evasive behaviour Holmes realised that there was only one thing to do. In order to raise the money for the boat fare he sold his pony for 150 rupees and set off for the palace in Rangoon. Surprised by Holmes' unexpected appearance, LCT effected great pleasure at seeing him – 'but only promises so far!' Holmes records in his diary. He was not surprised to hear that the company scheme to raise capital in London had failed but he was surprised, on returning to see LCT the next day, to get his overdraft paid off, 1500 rupees in his pocket for expenses, and an assurance that he would be paid in full by the end of August. Nevertheless, anticipating the true outcome, Holmes went to see solicitors in Rangoon and prepared to sue the company for his salary if this final promise was not forthcoming.

He returned to Yenangyaung to quietly sit out the rest of August in the office, writing up reports in anticipation of the inevitable outcome at the end of the month, and spending as little money as possible. Maggie and Norman had returned from Kalaw, and on arriving home one afternoon, weary after finally instructing the solicitors to start proceedings to recover his salary, Holmes found Norman feverish with severe diarrhoea. The doctors diagnosed dysentery. During the following week Norman's condition fluctuated, with soaring temperatures followed by drugged sleeps as the doctors administered saline infusions and morphine. Days alternated with hope and despair. In the middle of it all Dudley Stamp arrived to tell Holmes that he too had resigned, but they turned him away unable to deal with any more problems than this terrifying threat to their child. They were reluctant to give him too much morphine, so Norman

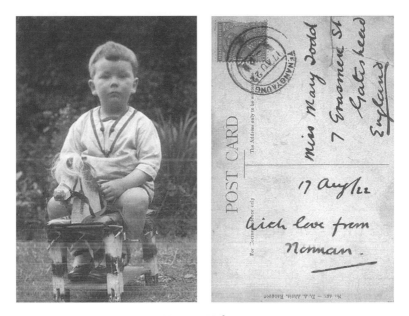

Norman Holmes.
'With love from Norman' was written by his father on a
postcard to a family friend in Gateshead, only three weeks
before Norman died.

struggled through two sleepless nights, with Maggie and Arthur
taking it in turns to sit by his bedside, before his condition
became so grave that he was finally taken to hospital at
Nyanghla.

The little wooden hospital had a consulting room, dispens-
ary and operating theatre on the ground floor, about a dozen
beds on the upper floor, two doctors and three nurses. Facilities
were extremely limited and had to cater for several hundred
British and Americans expatriates in the oil fields, and several
thousand Burmese and Indians. One had to be extremely ill to
be admitted. Norman was extremely ill, but however sick a
child was in those days, parents could not stay overnight, so for
three nights Arthur and Maggie returned home to try and sleep
while their three-year-old child fought for his life, alone in the

hospital. On Monday 11th September 1922, Norman's condition deteriorated further and the doctors warned that there was little left they could do. A slight recovery occurred following a hypertonic injection at four in the afternoon, but at six the doctors said goodbye, leaving them all in privacy. Norman passed peacefully away at 8.45 p.m., his distraught parents at his bedside.

The funeral was scheduled for five the following afternoon but Maggie was too sick to attend, having succumbed herself to an attack of dysentery. Friends and neighbours rallied round, some staying with Maggie, some going with Arthur to the funeral. It was a sad, sad day. Six weeks later Arthur and Maggie sailed for England leaving Norman behind in Burma; a wooden cross and a bunch of flowers on his grave.

Durham Days

*Never, never let them persuade you that things are too
difficult or impossible.*

Douglas Bader

On a bitter November day, Arthur and Maggie arrived back in
Gateshead on what should have been Norman's fourth birthday.
They went to stay with Maggie's parents having no money,
nowhere to live and no child to brighten their days. On the
journey home they had instinctively looked for Norman, expecting to see him come running. Habitually they would turn to see
what he was up to; look up in response to another child's cry,
or reach down for his hand that was no longer held out to them.
Then remembrance and grief would sweep over them. Over the
following months they wished over and over again that they had
left Burma at the same time as Stanley Hunter. Had they done
so their son would still be alive. But Holmes had taken his role
as manager very seriously and had tried hard to obtain the wages
owed to Yomah Oil's employees. He had hung on as long as he
could. Too long. Now they both bottled up their grief; Arthur in
particular was unable to speak to anyone of Norman's death,
not even his oldest friend Bob Lawson.

By the time he had left Burma Holmes himself was owed
nearly a whole year's salary, which he tried to salvage
through solicitors in London, but LCT, by then on the verge of

bankruptcy, had gone into hiding. Exactly a year later, after a warrant for his arrest had been issued following an application in the High Court that he be judged insolvent, Lim Chin Tsong died of influenza. He seemed to have lost the will to carry on. Although LCT's eldest son took over the business, it was obvious to Holmes that he was never going to succeed in recovering his money, when even the great might of the Burmah Oil Company had failed in getting theirs.

Over the next few months Holmes tried desperately to get a job back in academic life, but the vacancies created by the war years were now filled, and people were trying hard to return to leading normal lives; there was a disinclination to change jobs. In fact employment of any kind was hard to find, and hungry people marched from Glasgow to London as the country headed towards serious financial difficulties. First Holmes applied to the Gateshead Education Authorities for a grant so that he could continue his studies working from home, but despite his impressive list of publications they turned him down because he was not attached to any academic institution. To raise a few pounds he occasionally gave piano recitals, an occupation he greatly enjoyed, but which was unlikely to ever provide him with much of an income. Eventually things got so desperate that they were forced to ask Maggie's parents to re-instate her allowance, which had lapsed when they went abroad and which her father, perhaps reasonably feeling that his duties towards his daughter had been discharged, was reluctant to resume. It became a matter of some embarrassment between the two families and while they haggled Maggie and Arthur, unable to finance a place of their own, went to live with his parents, who could ill afford to maintain them in their small house in Whitley Bay.

Ultimately, in the summer of 1923, Holmes hatched a scheme with a cousin of Maggie's, a Miss Graham, who was an expert on furs. Together, but presumably with her money, they opened a small shop in the centre of Newcastle, she as a 'Manufacturing Furrier' and he as a 'Trader in Oriental Crafts', selling Indian brass goods, Chinese and Japanese fancy work, porcelain and other knick-knacks from the Far East. The dark interior of the shop filled with glittering goods created an impression of Aladdin's Cave, something rather exotic and unfamiliar in 1920s Newcastle. Furthermore, their prices were high and in the months after Christmas 1923 business went very slack. Holmes and Miss Graham worried about their long term prospects. Times were hard in the industrial north east of England and few people had spare money for knick-knacks.

Although Arthur had quite enjoyed playing shop-keeper when times were busy, when they were quiet and no-one came in the shop for hours on end, he became incredibly bored and thought continuously about the research he had left behind, some four years previously. How were things progressing on the age of the Earth – and what about his long-term dream of developing a geological time scale? Had anything changed while he had been away, was anyone any closer to answering those questions? He tried to keep in touch with latest developments, but it was difficult when he was not able to contribute anything himself. Furthermore, Maggie, although now pregnant with their second child, was undoubtedly still mourning the death of Norman. With Arthur refusing to talk about Norman, and perhaps blaming herself for his loss, she felt isolated and lost. Insecure and lacking in self esteem, she became increasingly uncomfortable with the relationship building between Arthur and his business partner, making life difficult for him when he got home, wanting to know exactly what had 'gone on' during those hours when they were cloistered together in the shop with nothing much to do. Pressured on all sides, Arthur longed for the old days when he could bury himself in his work, and began to despair at ever getting back into university life.

But by 1924 his luck had finally changed. In February Maggie gave birth to another son, Geoffrey, which kept her very busy and helped to ease the pain of Norman's death, and in May an opportunity arose that was so unique it must have made Holmes wonder if, after all, there was someone up there looking out for him.

Durham University had been established in 1832, but for one reason or another – largely the competition from colleges in nearby Newcastle – science had not flourished, and until 1924 the range of subjects taught at Durham was largely confined to the arts, particularly those subjects associated with theology. Essentially it was a small private collegiate university benefiting from its close relations with the clergy and cathedral, whose members served on the Durham Colleges' Council and who held considerable influence over all university matters.

As a result of provisions made in the Education Acts of 1918 and 1921, which ensured that no person was debarred from education because they could not afford to pay, there was an increasing demand for teachers in the expanding schools. Responsibility for training these new teachers lay with local councils; thus a fruitful alliance developed in the early 1920s between Durham County Council and the Durham Colleges' Council. As a result, in 1924, science was reborn at Durham University. Four new departments were created, a new building was constructed to house them, and seven new appointments were made – a professor, reader and lecturer in chemistry; a professor and lecturer in physics, and a reader in each of botany and geology.

Holmes was amongst eighteen applicants for the single geology post, and Bob Lawson amongst fifteen applicants for the professorship in physics. They were both placed on short-lists of three and called to attend interviews on Thursday 29th May, at 2 o'clock in the afternoon. It must have been a strange reunion for the two men who had seen each other only occasionally since Holmes had returned from Burma, now that Lawson lived in Sheffield. Throughout their friendship Holmes had always been the more successful academically while Lawson played second fiddle, and yet Lawson was now the one with an academic job and applying for the post of professor, while, if he got it, Holmes would only be a lecturer.

A couple of days later Holmes was writing to his old friend at the British Museum: *'My Dear Dr. Prior, You will be glad to hear that I have got the Durham post and a Department of my own; and in the first place I want to thank you for your own share in supporting me. I am glad to feel settled again and am looking forward with great interest to the work of building up from the very start.'* The note of relief in this letter to Prior is almost audible. Holmes had been appointed as Head of the new Geology Department at Durham with himself as the only member of staff. Lawson did not get the professorship in physics and as it was not worth moving from Sheffield unless it was upwards, he stayed where he was. In fact he stayed there for the rest of his working life, never progressing very much further. Although the two of them did collaborate on several more projects over the years, it was always Holmes who had the ideas and Lawson, with his mathematical ability, who helped implement them. It is tantalising to speculate on how strong the team might have become if Lawson had also got the job at Durham.

Holmes wound up his affairs at the shop, although Miss Graham continued trading for several more years, and spent the following months gathering up teaching collections from museums around the country and generally preparing for the arrival of students. On the 2nd of October, 1924, the new science department at Durham University was officially opened by the physicist Sir William Bragg (in fact Sir Ernest Rutherford had been the Council's first choice, but he was not available). Thus the new Science Building was ready for its first intake of fourteen students who were scattered in small groups, some singly, across the four departments.

College life at Durham in 1924 was still a very traditional and British affair even though there was an increasing intake of students on scholarships. Student 'rooms' were either a bedroom and a sitting room or, for those that did not have both, a room with a curtained-off alcove known as the 'horse box' which contained the bed and wash-stand. The room was heated by an open coal fire and each morning a 'gyp' would come in before breakfast to lay paper, sticks and coal in the fireplace and to make sure the coal bucket was full. He then went on to serve breakfast in hall. After breakfast 'bedders' came in to make the bed, clean the room and bring fresh water for the wash stand. Three meals a day were provided in hall, and every evening the formal wearing of gowns and black ties was required at dinner. Beer or cider could be ordered on 'battels', an account kept for provisions from the kitchen, but wine was for special occasions.

For offences against the recognised practices of the college the Senior Man had the power to 'sconce' any amount up to ten shillings, a considerable sum, and failure to show a pass when entering the college gate after it was closed at nine o'clock incurred a substantial 'sconce'. Ritual initiation rites for first year students were standard procedure and generally terrified the unprepared: on the second night of term a group of second and third year students would visit the first years in the small hours of the morning, get them out of bed and make them take off their pyjama jackets. They were then stood on a table and made to answer intimidating questions while a slimy red liquid, looking much like blood but was only red ink, was poured over them. In the morning the poor boy would probably find that his trousers had been hidden so that he would be forced to go to breakfast in his pyjama bottoms, whereupon he would see his trousers hanging on a tree or some place else that was equally difficult to reach.

Surprisingly, considering the religious connections of the university, daily attendance at chapel was not compulsory, and men and women undergraduates mixed freely except in the more regulated activities such as sport. Most of the students were training to be teachers, in fact, if scholarship students did not go on to this profession they would have to pay back their grant at the rate of £25 a year for the first ten years of their working life. Geology was not a major option in the first year, although the majority of chemistry or physics students took it as an auxiliary course. These first year geology classes became increasingly popular as Dr Holmes' reputation for being an inspired lecturer spread throughout the university. He had a lucid and exciting way of putting his subject across to the uninitiated that never failed to motivate his audience and, because he kept so up to date with current developments, they had the thrill of living on the frontiers of discovery. In particular he would give them regular updates on the age of the Earth, announcing with exaggerated gravity 'Today the age of the Earth is . . .'. He even attracted art students to come and listen, just for the pleasure of hearing him talk.

In addition, Holmes was considered a very 'fair' man by his students. As part of their final degree exams all students, regardless of the subject they were studying, had an examination in scripture knowledge that was set by the Bishop, but marked by the Head of Department. One year the question was, 'What significance do you attach to the minor prophets?'. When a student simply answered 'NONE', Holmes gave him full marks! Not because he was an atheist himself, but because he considered that the student had answered the question he had been set, and that the examiner was at fault for not setting the question he wanted answered. But Holmes was not only a good teacher, he was also kind and caring towards his students. Genuinely interested in what they wanted to do, he helped them wherever he could. As a result, some gave up their intended first subject and went on to specialise in geology in their second year, although in the early years there was usually only one or two a year of these privileged geology students.

The majority of these honours graduates went on to serve in geological surveys around the world, and the first of these, A. E. Phaup, was no exception. Phaup graduated in 1928, whereupon he saw an advertisement for a job as geologist on the Geological Survey of Southern Rhodesia. Soon after applying he was interviewed in London and immediately offered the job, never thinking to ask why the post had become vacant. On arrival in Salisbury he was handed a motor car (he had never learnt to drive), told to take on 'a dozen black men' and get straight out into the field. Finally curious as to what all the rush was about, on arrival at his destination he enquired what had happened to his predecessor. It transpired he had been eaten by a lion!

Holmes' second student was Kingsley Dunham, who eventually became head of the Geological Survey in Britain. Originally Dunham had gone to Durham to study chemistry and had taken geology as an auxiliary subject. At the age of eighty-eight Sir Kingsley recalled with great clarity the excitement of hearing Holmes lecture, and how it influenced the decision of his life:

To say that I was fascinated by his first year lectures is a great understatement: I thought the whole subject so exciting and it was the really delightful lectures by Holmes to his quite sizeable first year class that first introduced me to the subject. After the results of the auxiliary exams were posted I thought I would just look into Geology and say goodbye to Hopkins [by then Holmes was Professor, and Hopkins was a lecturer]. So I went in and he was sitting there at his desk. He looked at me and said 'Dunham, you did alright in that exam. Why don't you change over to geology and get into a real subject?' I was immediately very struck with this because I had enjoyed Holmes' lectures very much more than those by Dr. Walters who had given the physics lectures, or the reader in chemistry who had given most of the chemical ones. So I thought well, at least I know that by far the best of the three lecturers is Holmes, I may as well take a shot at this. I was also very conscious of the fact that none of my contemporaries in college either knew what geology was or had any idea of doing any such thing, so at least it was a chance to be different. The people I associated with were theologians, people reading honours in English or history and one mathematician. Those were my friends; that was the gang that used to meet on Saturday nights and drink coffee right through the night and argue about philosophical subjects. Great days.

Therefore, at the end of my first year, and with the agreement of all concerned, I gave up my candidacy for chemistry in favour of geology. I found myself the sole honours candidate of my year and it is not difficult to imagine the great advantages of individual tuition from men of the standing of Holmes. For the first time I began to find that I was being treated not as a schoolboy or a student, but as a man, and it was immensely thrilling to hear the gossip of the world of geology from the men who were involved in it. Holmes talked to me as though I were a colleague. This is marvellous for a student and it has lived in my memory ever since.

I completed my first degree in June 1930 gaining a first, indeed, any less a performance would have been an ungracious act, having regard to the excellent preparations given by Holmes and Hopkins, and by the time I got through my PhD with Holmes I knew a very great deal about the world of geology. Not just the subject, but also the people. No one could have been be more charming or pleasant.

Twenty years later Dunham was to tread in Holmes' footsteps as Professor of Geology at Durham.

While Holmes had been away in Burma the same old arguments had rumbled on with respect to the age of the Earth. At a British Association meeting in 1921 an impressive array of geologists, physicists and astronomers participated in a discussion that attempted, but failed, to bring into harmony the wide variance in time readings between 'hour-glass' methods and radioactivity methods, which still had to be reconciled with each other. Strutt, in place of Holmes, again tried to lay the spectre of Kelvin who still rose to haunt the assembly, and again put forward the arguments in favour of radiometric methods. William Sollas, Professor of Geology at Oxford, seemed overwhelmed at the amount of time now available to geologists after the paucity offered by Kelvin: 'the geologist who had before been bankrupt in time now found himself suddenly transformed into a capitalist with more millions in the bank than he knew how to dispose of but, perhaps understandably, he still urged caution and heeded geologists to substantiate the time being offered by the physicists 'before committing themselves to the reconstruction of their science'. But the die-hards were slowly growing smaller in their numbers and the meeting ended with a general acceptance that the age of the Earth was around 1500 million years.

The following year, in 1922, a similar meeting was held in America, sponsored by the American Philosophical Society, with speakers who spoke from the view point of a geologist, a palaeontologist, an astronomer and a physicist. The problems discussed and the conclusions reached were much the same as their British colleagues, but in particular the geologist, Thomas Chamberlin, took his colleagues to task for adhering to an age of 100 million years. The butt of his comments were specifically directed at John Joly and his refusal to renounce his ideas on an age of the Earth based on salinity in the oceans. So, although little new work had been done on age dating while Holmes was away, the tide of opinion was gradually turning and on returning to academic life in 1924 after a four year absence he noted *'there is now a marked change in opinion in favour of the longer estimates'.*

Finally, twenty years after Rutherford had dated the first mineral by radioactivity and twenty-five years since Kelvin's last publication advocating a 20 million year old Earth, there seemed to be a general acceptance that ages measured by radioactive methods were at least of the right order of magni-tude, and that the age of the Earth was to be measured in thousands rather than hundreds of millions of years. It was now up to exponents of the hour-glass methods to reconcile their ages with the radiometric dates, rather than the other way round. However, there were still many problems and a few individuals who persisted in sticking to the old methods. John Joly was one.

In 1925 Joly published a book on The Geological Age of the Earth, which favoured an age of between 160 and 240 million years to the Base Cambrian. This he deduced from his own method for measuring the age of formation of the oceans, which he still calculated to be 80–100 million years, and which he now supported by radiometric dates determined from thorium-bearing minerals rather than uranium minerals. Having been forced to abandon the helium dates he had once so favoured, and having lost the argument that uranium had decayed more rapidly in the geological past (and therefore the Earth was not as old as uranium dates indicated), he now homed in on thorium dates because they were often lower than those determined from uranium and so gave ages more in keeping with his age of the oceans. Although Joly was an established and well respected member of the geological community, Holmes did not hesitate to show him the error of his ways. In a polite but damning review of Joly's book Holmes again disposed of the sodium method – as he had done more than twelve years previously – by reviewing the evidence, or lack of it, and stating with some authority that: *'At least it is certain that the sodium method cannot at present be considered as making any serious contributions to the problem'.*

He then went on to show how inconsistent the thorium results were. Unlike those from uranium which showed a consistent correlation between lead and uranium, thorium–lead ratios were all over the place. Holmes thought that some of the lead isotope resulting from the decay of thorium must be escaping and, like the helium results two decades earlier, were resulting in values that were much too low. But the criticism of Joly was done in a gentlemanly manner and Holmes even ticked off another reviewer for having been discourteous to Joly – not everyone was as patient as Holmes.

At the other end of the scale Henry Russell, an American Professor of Astronomy, had determined that the age of the Earth lay between 2000 and 8000 million years by assuming that the total amount of lead in the Earth's crust was all produced from the decay of uranium. Clearly this was not so. Revising Russell's figures with what Holmes considered to be more realistic values for the amount of lead and uranium in the Earth's crust, Holmes brought Russell's age down to 3200 million but then himself turned reactionary. He stated emphatically *it is clear that the Earth . . . cannot have existed for so long as 3,200 million years'* because, he argued, all the ages from the oldest minerals so far dated fell largely in the 1000–1100 million years bracket. The oldest was only 1525 million years, therefore *'the frequently quoted age of the Earth, 1600 million years, thus appears to be of the right order.'*

It was problems like those posed by Joly at one end of the scale and Russell at the other end that caused Holmes to turn once again to his dream of constructing a geological time scale. If only radiometric dates could be determined on common rocks of a known geological age then all this uncertainty would disappear. Once a framework was established it would only be necessary to fill in the gaps as more and more data became available. But how to do it? What technique would prove reliable and easy enough to become routine? Minerals in the common rocks contained so little uranium that chemical determination of the tiny amounts of lead produced by radioactive decay was impossible. But even if it had been possible, the necessity for atomic weight determinations to then distinguish between the various lead isotopes was still so laborious that it was a major deterrent to the progress of uranium–lead dating. The method continued to languish.

But then in 1928 an opportunity suddenly presented itself. Fritz Paneth who had worked with Bob Lawson at the Vienna Institute of Radium during the First World War and who was now Professor of Chemistry in Berlin, had developed very precise techniques for measuring minute amounts of helium. Helium was ubiquitous – for each uranium atom that decayed, eight helium atoms were produced, so large concentrations of helium were generated over geological time, even in the common rocks. The problem here was not the shortage of uranium, but the surplus of helium which either could not be contained in the mineral over long periods of geological time, or was lost in the preparation process. But Paneth considered that precisely because the concentrations of uranium in common minerals (those not normally selected for age dating) were so low, the much smaller volumes of helium produced would have a good chance of being retained in the mineral. Holmes seized the initiative and arranged to have some common rocks analysed in Paneth's lab in Berlin.

Samples were taken from two famous rocks from northern England which were known to be of very different geological ages: the Whin Sill was believed to have a very late Carboniferous age, while the Cleveland Dyke was considered to be middle to early Tertiary. When Paneth sent the helium results back the Whin Sill gave 182 million years while the Cleveland Dyke gave 26 million years, values Holmes considered 'to be in excellent agreement with the geological evidence'.

The ideal test of these results would have been to compare the helium ages with the more reliable lead ages analysed on the same rocks, but it was precisely because these two rocks contained so little lead that they could not be dated by the lead method. Controls therefore had to be found in lead ages from other rocks believed to be of the same geological age. The problem was of course, that very few reliable lead ages were available anywhere in the world, so they had to make do with what ever they could get.

Arthur, Maggie and Geoffrey Holmes, in Durham *c.* 1930.

Today we know that the Whin Sill does indeed have a very late Carboniferous age but its radiometric date is around 295 million years, as opposed to the 182 million years measured by Paneth. Similarly, although the Cleveland dyke is indeed early Tertiary, its radiometric age is about 60, not 26, million years. The huge discrepancy in today's radiometric ages with Paneth's helium results well illustrates not only how poor the helium analyses were, but also how badly constrained the geological ages of most igneous rocks were at that time. Like the Whin Sill and the Cleveland Dyke, the control rocks were believed to be, respectively, Late Carboniferous and Early Tertiary, but clearly they were not. These significant geological errors facilitated the acceptance of wildly inaccurate radiometric ages being assigned to these geological periods.

But in 1929 Holmes was not aware of the poor geological constraints on the control rocks. As far as he was concerned, both of Paneth's helium results were only 10 million years younger than the uranium–lead determined ages for those geological periods. So strong was the desire to find a successful dating technique, he convinced himself that although the helium results were '*slightly low*', they concurred '*quite satisfactorily with the scanty results based on lead ages*'. Despite the historical problems with helium, Holmes could not suppress his excitement: '*there is now available a practical means . . . of building up a geological time-scale which, checked by a few reliable lead-ratios here and there, should become far more detailed than could ever be realised by means of the lead method alone.*'

At long last his dream of constructing a geological time scale really seemed to be within the bounds of possibility.

The Ardnamurchan Affair

Unfortunately I have so many irons in the fire at the
moment that there is practically no fire.

Groucho Marx

Ardnamurchan, on the south west coast of Scotland, is the
remains of an ancient volcano. Its unique rock formations open
a window into the interior of the Earth and provide geologists
with a singular opportunity to observe the processes, now frozen
in time, that occur deep within the crust and which have been
slowly brought to the surface over the last sixty million years.
In 1930, when James Richey of the British Geological Survey
published his completed geological map of Ardnamurchan
confirming that it was an ancient volcano, it caused much
interest amongst geologists. Consequently, the following year,
Richey agreed to run a field trip so that those interested could
look at the volcano in some detail and augment their understanding of the Earth's interior.

Kingsley Dunham drove his professor all the way from
Durham to Ardnamurchan in his two-seater Morris Cowley.
Maggie never accompanied Arthur on these trips because their
son Geoffrey was still only small and needed someone at home
to look after him, but as the new fashion of the time was to grow
tomatoes, the students joked that the real reason she stayed
behind was to water the precious tomato plants. The truth was

that Maggie did not enjoy university life: the lunch parties, the afternoon calls, and the expectations of a professor's wife did not sit well with her. Without a university education herself she felt out of her depth in this academic environment, unable to converse at their level, and she greatly missed having her family close by. Consequently, during their seven years in Durham, her relationship with Arthur had become very strained. Although they 'kept up appearances' as one did in those days, the marriage was not a happy one any more.

When Holmes and Dunham arrived at the hotel in Ardnamurchan it was to find a large group of people gathered from geology departments all over Britain. Many of them were either already, or were to become, great names in British geology, and amongst their number was Doris Reynolds, one of the very few women active in geology at that time. Doris was a lecturer at University College in London and already making quite a name for herself in petrology – the study of the origin and composition of rocks. An extremely boisterous and vocal woman in her early thirties with a large face and unruly hair tucked under a beret, she had strong opinions on absolutely everything, especially geology. She polarised people into loving or hating her, and her quick intelligence did not suffer fools.

Doris would have known intimately the published works of Arthur Holmes, particularly the two books he had written on petrology, the second editions of which had just been published, and his work on the volcanoes of Mozambique, which were of the same age geologically as Ardnamurchan. Holmes' opinion on many of the features they observed on the volcano would have been sought with some deference by Richey, thus making Holmes one of the leading figures in the group. Doris too would have been unable to resist vocalising her opinions on what they saw, so inevitably she and Holmes would have fallen into conversation discussing the merits of Richey's interpretation.

Back at the hotel later that evening, there was little to do in such a remote place but talk, and when geologists get together, they tend to talk about geology. One of the great geological debates of the day was the exciting new concept of 'continental drift' and it inevitably became the topic of conversation.

As early as 1620 Sir Francis Bacon had noted how good the 'fit' of South America against West Africa would be if they were placed side by side, eliminating the Atlantic ocean in between. Rather like two pieces of a poorly made jigsaw there were some places where they overlapped, and some where there was a gap, but overall the fit was remarkable. This 'fit' continued to fascinate and baffle people for centuries until in 1912 Alfred Wegener, a German mathematician and astronomer before he became a meteorologist, came up with the extraordinary suggestion that South America and West Africa had once been joined together. Not only that but he also postulated they were part of one united super-continent, which Wegener called 'Pangaea', that had contained the precursors of all the continents we see today. Wegener argued that during the Mesozoic, Pangaea had developed numerous fractures and the resulting continents slowly drifted apart until Cretaceous times when South America and Africa fully separated and the South Atlantic opened up between them. Wegener also recognised that North America separated from Europe much later on; that the same was true of the break between South America and Antarctica, and that Australia had detached itself from Antarctica as recently as the Eocene.

For many years it had been recognised that the rock formations and the fossils found on the coasts of both Brazil and West Africa were identical, although the two countries were now 5000 miles apart on opposite sides of the Atlantic. Fantastic theories were postulated to explain these phenomena, the most widely accepted being that of a 5000 mile land bridge which had once linked the two countries but which had since sunk without trace into the ocean floor. Citing Panama as an example of a land bridge which now links North and South America, adherents to the 'Atlantic land bridge theory' argued that animals whose fossil remains were found on both sides of the Atlantic could have walked from one side to the other, taking with them the seeds of plants and trees, the fossils of which were also found on both sides. It did not seem such an unreasonable idea but for the huge distances involved, and the extraordinary shape of the two continents.

One of Wegener's strongest arguments in favour of his theory was that when the continents were reconstructed in this way then an ancestral equator could be traced right across Pangaea from the fossils typical of a tropical environment, such as those which had prevailed during the formation of the coal fields of Britain and many other countries during the Carboniferous. The inference from this was that Britain had lain close to the equator during the Carboniferous, as part of Pangaea, and had been drifting northwards ever since Pangaea split up. In a similar way, evidence for past glaciations of the same age was also found on many continents that were now thousands of miles from the poles and each other, giving further strength to the Wegener hypothesis. This remarkable theory seemed to explain so many geological conundrums, but the concept that continents could drift helplessly around the world covering many thousands of miles, when they seemed to be so solid and stationary, was so preposterous that the majority of geologists just could not accept it.

Apart from the obvious difficulties it posed, the main obstacle to accepting the theory was the lack of a mechanism which could physically move continents large distances. Wegener visualised the continentals blocks as slabs of granite and gneiss which, being rich in the lighter elements, floated in a sub-stratum (now known as the mantle) containing the heavier elements, but he was unable to come up with any convincing driving force that would move these slabs around. As one opponent succinctly put it: 'We are invited to think of a continental mass under almost no horizontal forces ploughing its way steadily through an ocean floor, which resists its advance enough to squeeze up folded mountains against nothing at all'.

Admittedly, it was a difficult problem to explain, but Holmes had given it much thought over many years. One of his colleagues while at Imperial College, and with whom he became very friendly, was John William Evans, a remarkable man who, in 1914, was appointed as a demonstrator in the geology department at the advanced age of fifty-seven. Despite being educated in a legal career and called to the bar in 1878, Evans had succumbed to the attractions of geology and had gone on to study it at the Royal College of Science. Following many years of geological adventures in the remote parts of Brazil, India and Bolivia, he returned to his old college at the outbreak of war, being too old to be called up. A complete eccentric in the true British sense, he lectured in an overcoat with the collar turned up, a woollen scarf around his neck and a black Homburg hat perched precariously on his hairless head. The archetypal absent-minded professor, he spent hours looking for mislaid objects and was never seen without several dictionaries and grammars of various languages under his arm, available for study at odd moments in buses or trains. He was a man of extraordinary wide interests and according to P. G. H. Boswell, the geology professor who succeeded Watts, Evans was closer to a genius than any other geologist he had ever met.

The subject of continental drift was dear to Evans' heart. Fluent in several languages, he had been able to read Wegener in the original German and became an early convert to his theories. Having a great respect for Holmes and his work on radioactivity, he undoubtedly discussed with Holmes the merits and flaws of continental drift long before Wegener's translation into English, to which Evans wrote the foreword. Tossing backwards and forwards ideas for a mechanism that could drive the continents through the crust, Holmes had become steeped in the controversy long before he went to Burma. When he came back the third edition of Wegener's book had just become available in English so, by the Ardnamurchan meeting almost ten years later, and nearly twenty years since Wegener had first proposed his ideas, the controversy was in full flood.

As we have seen, geologists are frequently not very receptive to new ideas. Years later, at the age of eighty, Doris Reynolds succinctly summed up the situation:

Geology is always like this, very slow moving. When a new geological discovery or suggestion is made it is quite quick if it is noticed in 20 years, and may take 50 to 100 or more. Then dogmas form obstructions.

Indeed, for fifty years dogma formed obstructions to continental drift, and only a few enlightened individuals recognised early on that it was the only way so many geological phenomena could be explained. Arthur Holmes had long been part of that small group.

While work on the uranium–lead dating techniques was in abeyance, Holmes was looking for some other problem to get his teeth into. With his profound understanding of radioactivity – the amount of heat it generated within the Earth, and the enormous time it bestowed on geology for infinitely slow processes – he found himself placed in a unique position to formulate a mechanism for continental drift. So as soon as he had established his teaching routine at Durham he seriously started to formulate his ideas, and in December 1927 he read a groundbreaking paper to the Edinburgh Geological Society. In it he proposed that differential heating of the Earth's interior, generated by the decay of uranium and other radioactive elements, caused convection in the substratum, on which the continents floated, rather like icebergs in the sea. Although the substratum was essentially considered to be solid, Holmes believed that given enough time, and of all people Holmes knew that there was enough time, it actually behaved like a very thick liquid. (Roman glass, for example, today shows evidence of having 'flowed' during the last 2000 years.) The differential heating caused convection cells to form, rising in some places and descending in others, behaving in much the same way as treacle would if heated in a saucepan on the stove. As hot material reached the top of a convecting cell beneath a continent it would travel horizontally for some distance before cooling and descending again. As it travelled horizontally it would produce a force that was sufficient to drag the continents sideways, and so the continents were very slowly pulled apart allowing the substratum to rise up and take their place in the ocean floor.

Leading geologists, particularly the Americans, were exasperated by these ideas, as this letter from William Bowie illustrates:

Holmes brings out a new thought which is even more impossible than Wegener's. That is that the submerged ridge through the Atlantic Ocean is the place at which North and South America separated from Europe and Africa, the latter two continents drifting eastward and the Americas drifting westward. I do not see how the same force, operating to send one

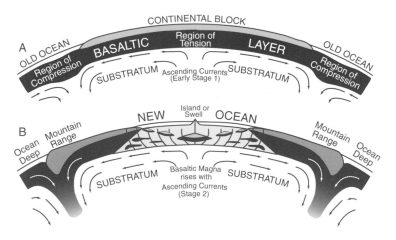

Continental drift.

These diagrams illustrate what Holmes cautiously called 'a purely hypothetical mechanism for "engineering" continental drift'. He described how in A the sub-crustal currents are in the early part of the convection cycle, while in B 'the currents have become sufficiently vigorous to drag the two halves of the original continent apart, with consequent mountain building in front where the currents are descending, and ocean floor development on the site of the gap, where the currents are ascending'.

mass westward, could make another go eastward. I believe that we need to apply elementary physics and mechanics to the continental drift problem in order to show how impossible drifting would be.

Another particularly vocal opponent was the eminent mathematician and physicist, Harold Jeffreys, who argued that continental drift was 'out of the question' as no force was adequate to move continental slabs over the surface of the globe.

Despite having given his first public lecture on continental drift in 1927, Holmes' seminal paper on the topic was not published until 1931, just a couple of months before the Ardnamurchan trip, so inevitably he became the obvious focus of attention as the beer slipped down and the debate heated up. Once again Holmes found himself in a minority of one. Attacked from all sides he was fortunately well versed in holding his ground when necessary and tried to catch his opponents off-guard by apparently agreeing with them:

'There is of course a very deep rooted prejudice against continental drift, moreover Wegener's book is rather calculated to provoke hostile reactions. It is easy to prove Wegener wrong over and over again, but proving Wegener wrong is by no means equivalent to disposing of continental drift.'

'So what evidence do you have that it occurred?'

'To my mind the distribution of climates during the Upper Carboniferous constitutes conclusive proof that continental drift operated on a very extensive scale. The occurrence of tillites [glacial deposits] of the same geological age in South America, South and Central Africa, India and Australia presents us with a hopeless riddle unless we assume that the glaciated lands were then grouped together near the South Pole.'

These arguments were difficult to refute and could not be explained by land bridges.

'But what about convection currents, where is the evidence for those?'

This was harder to be persuasive about, but data from earthquakes was beginning to lend support:

'The distribution of earthquakes occur near the shores of the Pacific, in the East and West Indies, in the mid-Atlantic and under the Himalayas. If these deep-seated disturbances are not in some way a result of currents operating far down in the zone of flowage, it is difficult to conceive a mechanism that could be responsible for them.'

'But surely all this is just speculation?'

'Discussion and speculation are justified if they do no more than stimulate a search for the more elusive pieces of the jigsaw puzzle we are doing our best to put together. If the suspected currents in the substratum form part of the completed picture, we must look carefully for possible evidence of their existence.'

And so the discussion continued late into the evening – Holmes erecting *'wickets to be bowled at'*, something he was fond of doing and which he hoped would stimulate new ideas and ways of thinking about the problems, and others trying to knock them down. Holmes himself had little doubt that his theories were close to the truth, but it was to be another thirty five years before his and Wegener's ideas were shown to be fundamentally correct, and even longer before Holmes himself was given full credit for them. Not until 1985 did Robert Muir Wood write:

As it took Wegener, the dreamer, to conclude that the continents were drifting, so it took Holmes, the critic of geology, to place continental drift for the first time on a scientific foundation. No longer were the continents to be driven by forces weaker than the slightest breeze. There was a mighty engine deep in the Earth, powered by radioactivity.

A handsome man in his early forties with thick dark hair brushed back off his high forehead; well dressed in tweed plus-fours, shirt, tie and matching tweed cap even when out in the field, Arthur created quite an impression on Doris Reynolds that day. They stayed up late into the night discussing geology and the following day were almost inseparable as they walked around Ardnamurchan. That evening, not wishing to get into the heated debate of the previous night, Holmes and his student Dunham amused everyone by playing four-hand duets on the hotel's piano. The professor and his most junior demonstrator thumping out tunes on a piano that was out of tune and had a couple of keys missing, put everyone in the party mood. Musical Chairs caused much hilarity as grown men fought with one another to get seated. Doris won musical statues to cries of 'favouritism' at Arthur who was both pianist and judge, then someone suggested Cat's Cradle. Here two people are placed back to back in the centre of a large circle of people and a piece of string is passed through their hands. The ends of the string are held by the circle which revolves around them eventually tying them up in a cat's cradle from which the two then endeavour to extricate themselves. The first people to be called upon were Doris and Arthur. Their interest in each other had not gone unnoticed by the group, and somebody was feeling mischievous. As they stood back to back with everybody looking on Arthur felt a good deal of embarrassment, but also highly flattered at having been singled out from all the other men on the trip by this brilliant woman. Doris on the other hand was laughing, not the least self conscious, and enjoying every minute of being tied up with Arthur. Even then, Kingsley Dunham recognised it as a symbolic moment.

By the end of the ten-day trip Arthur and Doris were in love. Fired by an unbridled passion, their mutual love of geology, it was a true meeting of minds, their individual interests in the science dovetailing perfectly. They talked geology incessantly, and on returning to their respective departments they continued their geological discussions by letter. They must have occasionally seen each other when Holmes visited London for Geological Society Meetings, for he records his grateful thanks to *'my friend Miss Doris L. Reynolds . . . for her generous co-*

Doris Reynolds at about the time she met Arthur Holmes.

operation' in drawing the figures for some of his papers. When a friend questioned Doris as to why she was so often acknowledged in his papers of the time, when she had actually done very little to assist, she replied simply, 'He was in love with me!'. Within a year the relationship had blossomed into an affair, as Doris' father found out when he accidentally opened a letter to her from Holmes. His profound disapproval was voiced in his silence on the matter.

Arthur Holmes, Professor of Geology at Durham University, at
about the time he met Doris Reynolds.

In the early 1930s Holmes was rising towards the peak of his
career. His reputation as a deep thinker on geological problems
and, as the American geologist Reginald Daly described him,

'one of the few English geologists with ideas on the grand scale', led to requests for him to lecture around the world. In 1930 he was an exchange professor in Switzerland, in 1931 he toured Finland extensively, and in 1932 was invited to give the prestigious Lowell Lecture Series in the United States. In the same three-year period he published no fewer than twenty-one papers on his research, which he typed himself on an upright typewriter, and updated the second editions of his two books on petrology. As well as fulfilling his normal teaching duties he served on several committees, including one to investigate the age of the Earth. By anyone's standards, particularly in the days before air travel and computers when everything progressed much more slowly, Holmes was a busy man. With only one other lecturer to help him in the department, it was not unreasonable that he put in a request for extra assistance.

In January 1933, at a meeting of the Joint Board Appointed to Administer the Departments of Science and Education in Durham, the head of the Science department proposed, 'That the time has arrived for considering the necessity of increasing the staff of the Geological Department'. He submitted details indicating the excessive amount of work falling upon Professor Holmes and Dr Hopkins and suggested an addition to the staff of a lecturer in geology. Within a month permission to appoint the lecturer was forthcoming and a special sub-committee was set up to advertise the post, consider applications and recommend a selection for interview. In the latter connection the sub-committee was empowered to invite Professor Holmes to act in consultation. No fewer than thirty-seven people applied for the post, and amongst them was Doris Reynolds. Three were invited for interview, and amongst them was Doris Reynolds. With Holmes advising the selection committee it was hardly surprising that she got the job. On the 17th of October 1933 Doris Reynolds was installed on the opposite side of Arthur Holmes' large desk – ostensibly because there was no room for her elsewhere in the department.

During his trip to the United States in 1932 Holmes had discussed with his colleagues there the age dating results he had obtained using Paneth's new helium method that appeared to be so successful. On an earlier occasion he had put out a plea to American universities to help him with his dream of building a geological time scale, their wealthy research establishments being the only ones that could really afford to consider this *'Herculean task'*, so while in America, and with the positive helium results from Paneth to support his case, he raised the subject again. His reasoning was persuasive and it was agreed that William Urry, then working with Fritz Paneth in Berlin, would come to the United States and finally make Holmes' dream come true.

The samples were carefully selected from sites all over the United States, Canada and Europe and their geological ages ranged from the Cambrian right up to the top of the Tertiary; the crucial factor determining a sample's suitability being how well its position in the geological column could be constrained. Rocks would only be chosen if their geological age could be really well defined. The idea was to use basalt, the rock that flows from volcanoes as molten lava, because it was easier to assign a geological age to basalts than any other common igneous rock type. In addition basalts were widely distributed in space and time and so representatives of all ages were readily available.

When basalts are erupted from a volcano as a flow of lava, they may travel great distances (hundreds of kilometres) and as they do so they trap beneath them soil containing organic life that will eventually become fossilised – as Lyell discovered when he found the fossilised remains of shellfish beneath lavas while examining the volcano Mount Etna. Once the volcanic action has ceased and the lavas stop flowing, sediments will again be deposited on top of the basalts. In time these sediments, also containing living organisms, will eventually be turned into fossil-bearing rocks. In this way many basalts acquire well-defined fossilised brackets below and above the flows which can be used to assign a quite precise *geological* age to the basalt.

Working at the Massachusetts Institute for Technology (MIT), Urry took nearly four years to analyse the thirty-nine samples initially selected, but when Holmes finally saw the results he was over-joyed. He was just rewriting the third edition of his book on *The Age of the Earth* and in it he proudly compared Urry's helium ages with established uranium–lead ages. The correspondence was remarkable. Another wild miracle seemed to have occurred. Holmes wrote euphorically in his book:

. . . it is remarkable how consistently the age estimates fall into appropriate positions. That this stringent test of internal consistency is satisfactorily met must be regarded as the final proof that the ages calculated from lead and helium ratios are at least of the right order and that no serious error is anywhere involved.

Remember three steps forward, one step back? This time it was two steps back. As Holmes' book was going to press a problem was noticed by an assistant professor of physics, Robley Evans, who had recently come to MIT to develop yet another new method for determining helium ages. When Evans' new technique was sufficiently well established and giving consistent results, ages were determined on some of the same samples that Urry had used and the two sets were compared. Urry's ages were found to be significantly higher than those determined by Evans. After exhaustive investigations the problem was traced to an error in Urry's equipment which resulted in all his ages being consistently too high, and fortuitously in line with the uranium–lead ages. It was a devastating moment as they realised that five years of hard work would have to be discarded, and it was back to the drawing board. Helium was still leaking, even from these barely radioactive samples. Evans summed up the position: 'These general inconsistencies in helium age ratios indicated quite clearly that there was some fundamental failure

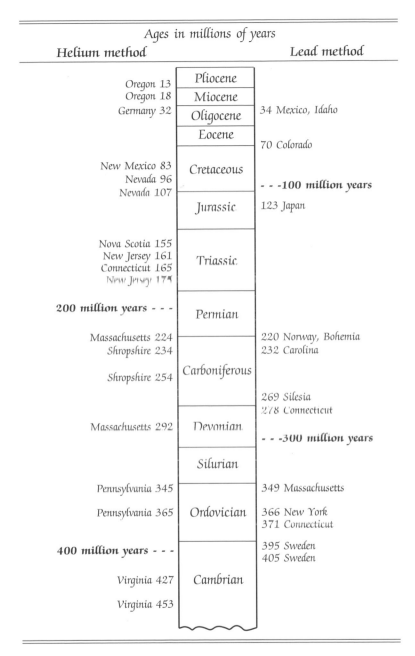

Helium method	Ages in millions of years	Lead method
Oregon 13	Pliocene	
Oregon 18	Miocene	
Germany 32	Oligocene	34 Mexico, Idaho
	Eocene	
		70 Colorado
New Mexico 83	Cretaceous	
Nevada 96		
Nevada 107		- - -100 million years
	Jurassic	123 Japan
Nova Scotia 155		
New Jersey 161	Triassic	
Connecticut 165		
New Jersey 175		
200 million years - - -	Permian	
Massachusetts 224		220 Norway, Bohemia
Shropshire 234		232 Carolina
Shropshire 254	Carboniferous	
		269 Silesia
		278 Connecticut
Massachusetts 292	Devonian	- - -300 million years
	Silurian	
Pennsylvania 345		349 Massachusetts
Pennsylvania 365	Ordovician	366 New York
		371 Connecticut
400 million years - - -		395 Sweden
		405 Sweden
Virginia 427	Cambrian	
Virginia 453		

William Urry's 1936 helium results
Holmes compared these with world-wide lead results in his 1937
edition of *Age of the Earth*.

in the helium method'. But it was too late for Holmes' book, which was already in print. Wise, many years after the event, he wrote: *'We have all been hopelessly wrong at one time or another, and it would ill behove us to hold the great pioneers in any less esteem because we have the advantage of techniques undreamt of by them.'* Was he perhaps, including himself amongst the 'great pioneers'?

Sitting opposite one another across the huge desk, surrounded by piles of books, Doris and Arthur were blissfully happy in each others company, but others were not so pleased with the new arrangement. Doris had taken over Holmes' teaching of the first year geology class which, with thirteen men and nine women, was the biggest Durham had ever seen. But although she was a very capable and sometimes even entertaining lecturer, she did not have the same ability to spellbind in the way that Holmes did. In addition, his research students complained that they could never see Holmes without Doris being present and having a say in the matter whether it involved her or not, so the close relationship they had built with him became diluted. Furthermore, tied to the cathedral as the University was, for both its financial and spiritual well-being, any improper behaviour was unlikely to be tolerated by the authorities. Whatever he might have wished, Holmes was still a married man with a young son, and he was treading on very dangerous ground. But he was in love and apparently oblivious to criticism.

Doris rented a house on her own, just across the university campus from the Holmes' residence, so she and Arthur would often be seen standing together at the bus stop outside the Science Laboratories talking geology. Arthur would see her home and then walk across the park to his house, sometimes quite late at night. Undoubtedly his long hours and frequent absences from home, even before he met Doris, contributed to his estrangement from Maggie. Frantic to make a success of the job, the memory of not having one still being a stark reality, Holmes had driven himself hard and was rarely at home. Maggie on the other hand disliked university life and the people around her. Lonely and with too much time on her hands to think and brood, she perhaps drove Arthur away, eventually to seek solace with Doris. It was a vicious circle. Furthermore, probably because he was away so much, Arthur never managed to establish quite the same bond with his son Geoffrey as had developed between father and son when Norman was born, so there was even less to tie Arthur to the home. Geoffrey had been born the year that Holmes started at Durham.

But Arthur was fundamentally a kind man and terribly torn between his love for Doris and concern for his wife and son. So husband and wife persevered with keeping up appearances and it seems highly unlikely that Arthur would ever have left his family for Doris. So he continued his illicit relationship with her for another five years, seeing her all day at work, sharing some evenings, and once a year taking their students on a field trip to Ireland together. It caused a great deal of gossip, but neither of them seemed to care.

Maggie had never been in very good health, suffering from asthma ever since she was a child, and always having a chesty cough. But early in 1938 it became clear that she was seriously ill. She became very thin and had difficulty in eating. Stomach cancer was diagnosed. The doctors decided to operate but even as she recuperated in a nursing home, Arthur and Doris were seen going to the cinema together in Newcastle where they thought they would escape prying eyes. Maggie returned home and recovered for a short while, but over the following months she grew weaker and weaker. Quietly, and perhaps one might even say conveniently, in September 1938 Maggie Holmes died.

Rewards and Retributions

Experience never misleads; what you are misled
by is only your judgement.

Leonardo da Vinci

Progress on dating the age of the Earth was slow. Years, even decades went by without any significant advance being made. But science is like that. What is often not realised when the breakthrough finally occurs is that for years previously a few individuals had been diligently working in the background, thinking and writing about the problems, quietly and persistently pursuing their goal. Arthur Holmes was one. Every few years he took it upon himself to write an article summarising the current state of play with regard to the age of the Earth. In simple and lucid language he explained to the scientist and layman alike the history of radioactivity, its application to dating minerals and the age of the Earth, and included any recent developments. Year after year he said much the same thing: Kelvin's arguments were shot down in flames, the 'hour-glass' methods were swept aside, and radioactivity emerged victorious. Slowly, bit by bit, this one-man campaign spread the word about the great antiquity of the age of the Earth.

He also continued to build up the database. As early as 1923 a committee had been set up in America for 'The Measurement of Geologic Time by Atomic Disintegration', its objectives being to collate and monitor all the dating of rocks being done around the world. In 1926 a sister committee on the age of the Earth was commissioned to address specifically the wide variance in time readings derived from the hour-glass methods and radioactivity. Despite *'declaring against committees'* Holmes was a founder member of both and the representative geologist, along with an astronomer, a physicist and a palaeontologist, on the latter.

In 1931 the Age of the Earth Committee published its findings in a book that addressed the age of the Earth from the sedimentalogical, the palaeontological, the astronomical and the radioactive points of view. Holmes' contribution on radioactivity and geological time filled three quarters of the book. He covered everything from the discovery of radioactivity to the latest ideas on quantum mechanics and the disintegration of atoms. Then, having scrutinised in detail every analysis of uranium, lead, thorium, radium and helium that had ever been published anywhere in the world, now amounting to many hundreds, Holmes determined an age for the Earth. It was a true labour of love to which he dedicated years of his life. During this period there were no wild miracles, no leaps of the imagination, no flashes of intuition, just continuous and rather monotonous, repetitive hard work. One by one he examined the analyses and *'After rejecting the lead ratios of all materials that for one reason or another do not fulfil the criteria of reliability'* (and all the helium analyses) he finally calculated the ages for thirty 'acceptable' uranium–lead ratios, standardising all the results. His conclusion was that *'No more definite statement can by made at present than that the age of the Earth exceeds 1460 million years, is probably not less than 1600 million years, and is probably much less than 3000 million years'*. As precision had increased and results were re-calculated, the age of the oldest mineral so far discovered had gone down slightly, but in twenty years of radiometric dating, the age of the Earth had not changed by one single year, only the error bars had got wider.

Helping him with this gargantuan task was his old childhood

friend Bob Lawson, whose mathematical ability Holmes still relied upon. A touching acknowledgement at the end of the book reveals the depth of their friendship:

Doctor Lawson and I have frequently collaborated during the last two decades, and the development of the subject owes more to his influence than his published contributions can indicate. Throughout the writing of this . . . I have had the great advantage of constant discussion with him. His experience as an investigator of radioactive phenomena has been freely placed at my disposal and my indebtedness to him is greater than can be formally expressed.

Even today we can feel the tremendous warmth and respect that existed between these two men.

Another, more conventional acknowledgement was to Dr Francis William Aston for providing Holmes with data on lead isotopes.

Back in 1910 Aston had been working at the Cavendish Laboratory in Cambridge. Like all the others there at that time he too had been carried along with the excitement of radioactivity and the new theories emerging about the atom. Thus when isotopes were discovered Aston helped develop the first mass spectrograph, forerunner of the mass spectrometer, an apparatus which separates isotopes according to their atomic weights by passing them through a magnetic field. Development of the mass spectrometer was to revolutionise dating methods. Today it is a highly sophisticated instrument that can analyse a sample for its uranium and lead isotopes (or almost any element you care to think of) in a few minutes, and tiny mass spectrometers are put on the Moon and Mars probes to test the composition of the atmosphere and the soil, giving almost instantaneous readings. But in Aston's day they were unwieldy, hand-built instruments requiring constant care and attention.

By 1922 Aston had been awarded the Nobel Prize for Chemistry, and by 1927 he had identified on his mass spectrograph the three known isotopes of lead previously recognised by their atomic weights. It therefore came as quite a shock when in 1929 some new results indicated that 'ordinary' lead could not, after all, be the lead isotope that had been around since the beginning of time, and that instead, it must result from the decay of a hitherto unknown isotope of uranium. Aston discussed the problem with Rutherford who agreed with his deductions, and went on to calculate that the new uranium isotope must have an isotopic number of 235. Indeed, it was uranium 235 which, because it represented less than one per cent of total uranium, had so far gone unnoticed. Wild miracle? Wild certainly, as uranium 235 unleashed on an unsuspecting world the opportunity for mass destruction. The miracle is, perhaps, that we still endure.

By estimating the rate at which uranium 235 decayed, which was much faster than uranium 238, and assuming that equal amounts of uranium 235 and 238 were present when the Earth first formed, Rutherford was able to determine the time it had taken to increase the ratio of the two uranium isotopes from zero to its present day value. The figure he arrived at, 3400 million years, was the first age of the Earth to be based on isotopic data from a mass spectrometer. Once again, Rutherford had got there first. But for some reason no-one seems to have taken this age very seriously and it was largely ignored. Perhaps because there were still many uncertainties – the exact decay rate of uranium 235 and hence the exact ratio of 235 to 238 – but more likely was the fact that it was just 'too old', even for those working in the field, to take on board.

So, 'ordinary' lead, lead 207, was even more ordinary than originally thought, and turned out to be just another decay product of uranium, albeit the newly discovered uranium 235 isotope. What then, was 'ordinary' lead? Did it exist at all? Well, yes it did. Holmes had already identified samples that contained

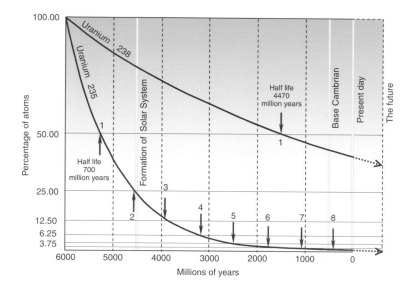

The decay rate of uranium 235, compared with uranium 238.
Today we believe that uranium formed in a supernova more than
6000 million years ago. This material was then inherited by the
solar system, of which the Earth is a part, around 4600 million
years ago. The production ratio of uranium 235 to 238 in a supernova is about 1.65, but in 1929 Rutherford assumed they were produced in equal amounts. From the diagram it can be seen that
uranium 235 decays much faster and so has passed through more
than six of its half lives by the time uranium 238 has decayed to
only half its original amount.

far too much lead, given the amount of uranium or thorium
present, so there must have been some there already when those
elements started to decay. But what was its isotope number? The
question remained unanswered for several years until, in 1936,
a young physicist was invited to join a team at Harvard University
to work with the latest and best designed mass spectrometer.
Although a physicist by background, Alfred Nier soon became
interested in the problems of measuring geological time, and
resolved to try and identify all the known isotopes of lead on his

new machine. Great efforts were made to ensure that all the equipment, built by hand, was free from contamination before the new analyses were made. When the results came through, as expected, there were the three lead isotopes already recognised by Aston, but right at the end of the spectrum a tiny blip could just be seen. Previously obscured by contamination, this minute spectrum of 'ordinary' lead was finally visible in this super-clean equipment, and identified itself as lead 204. The missing piece in the uranium–lead jigsaw had at last been found, twenty-five years after isotopes were first discovered.

Following these results, Nier wrote to Holmes whom he considered 'more than any other was central to ideas . . . in which lead isotope studies of ores and rocks played a crucial role, [such as] the construction of an absolute geologic time scale, ages of oldest crust and minerals [and] the age of the Earth'. Nier asked Holmes if he would like to collaborate on some analyses and, if so, to send Nier some rocks. But in the same way that the young Holmes had criticised Bertram Boltwood in print way back in 1911 for an omission in one of his papers, Nier had once supported critics of Holmes' theory on the origins of lead ore which, it turned out, was based on some faulty data. Holmes thanked Nier *'most cordially'* for sending him particulars of *'your most important work'*, but despite the fact that Nier only needed 10 milligrams of lead for a complete analysis and that it was possible to analyse a sample every two days, something Holmes must have dreamt of being able to do, Holmes replied: *'I have no specimens that it would be worth while to send you at the moment.'* A fit of pique perhaps? To be fair, Nier's letter did arrive shortly before Maggie died when she must have been gravely ill, so perhaps Holmes had other things on his mind at the time.

It was to be another six years before Holmes needed some data from Nier, whereupon a wonderful correspondence and collaboration developed between them that resulted in a major breakthrough for dating the age of the Earth. Unfortunately, during those six years demands on Nier's time and mass spectrometer had greatly increased and it became much harder for him to do work for Holmes. But such is the pace of science – dependent upon the whims of us mere mortals.

Only nine months after the death of Maggie Holmes, Arthur and Doris were married on the 30th of June 1939 at Durham Registry Office, and a small party gathered afterwards at their large new house. Like Maggie, Doris too had the advantage of modestly independent means from her family, and as a token of her love and affection she gave Arthur a Bechstein grand piano, which stood proudly displayed in the lounge. He gave her some beautiful jewellery. It was a small group who gathered to congratulate them, and conspicuously absent from the guest list are Bob Lawson and his wife Winifred. Did they perhaps disapprove, or had they known Maggie too well not to be invited? Given the endurance record of Arthur's friendship with Bob, it is unlikely that even disapproval would have kept Bob away, had he been invited. It seems more probable that Doris prevailed upon Arthur not to issue an invitation, and sadly there is no evidence of the two men ever collaborating again after the wedding.

As with Arthur's first marriage, within a few weeks of the ceremony a World War broke out, but within six months of that Arthur and Doris were embroiled in a war of their own. The formalisation of their relationship seems to have done nothing to appease the university authorities, who had been outraged by its previous illicit nature. On the contrary, they finally found the opportunity they had been waiting for to display their disapproval. At the end of the 1940 academic year Doris' contract as a lecturer was due for renewal. Normally, as long as her work was satisfactory, this would be a simple formality, but after a meeting held in the January the Committee on Re-Appointments reported to the Academic Board in the following terms, regarding Dr Reynolds' lectureship in the Geology Department:

The Committee agree that the work of Dr. D. Reynolds (Mrs. Holmes) in all respects justifies her re-appointment on the higher scale of salary. The Committee suggest that, before a recommendation is made, Council be asked whether or not they approve of the employment of husband and wife in the same Department.

The Academic Board accepted, without discussion, the Re-Appointments Committee's recommendation to refer the matter to the Durham Colleges' Council the following week. It came like a bombshell to the Holmes who luckily heard about the matter just in time. Arthur leapt to his wife's defence and wrote to Dr Applebey, Chairman of the Durham Colleges' Council:

I heard of it [the recommendation] with surprise and distaste. It seems to me both dangerous and unnecessary to raise the question in this general way, and I hope Council will take the view that they will raise no objection so long as the welfare of the Department concerned is in no way impaired.

I can assure you beyond all possibility of doubt that in this particular case the Department has been strengthened rather than weakened, since we both have now more time to devote to it and to the development of our subject by research than was formerly possible. No other change of any kind is, or will be, involved. My wife's career as a geologist is her dominant interest, no competing interest of any kind having been introduced by her marriage, and it would be a most unfair and cruel blow if Council passed any resolution that would lead automatically to her dismissal. Dr. Reynolds has an international reputation as one of the leading petrologists of this country and, I may add, she is the only member of the Science staff who has earned the degree of D.Sc. since the Laboratories were opened. Any man of similar achievement would already have been elected to a Professorship. If she had to leave, the Department would suffer a very grave loss, as it would be impossible to fill her place by anyone of comparable attainments, either as teacher or researcher.

When the Durham Colleges' Council met later that week, it passed the buck back to the Academic Board with a thinly veiled directive: 'It was agreed that, if the Board recommended the re-appointment of Dr Reynolds for an experimental period of one year, Council would agree.' The normal renewal period was for five years so it was a cleverly designed affront to the Holmes when, after seven years of lecturing in the Geology Department, Doris' contract was renewed for a single '*experimental period of one year*', as if she was on probation. It was obvious to the Holmes they were now being punished for their previous misdemeanours and that their position in the department had become untenable. It was time to look for another post where their history was not so well known.

Why does the Sun Shine?

The Earth seems to have been born from the star
spray of a stellar collision, but only in the realms of
imagination can her destiny be foreshadowed.

Arthur Holmes

During the Second World War, as the Blitz raged over Britain,
sixty thousand innocent civilians died and fifty thousand were
injured, either caught under a direct hit or trapped in a burning
building. Night after night the bombs rained down as more
than a million homes were destroyed, an immense amount of
damage was caused to industrial installations and many public
institutions were ravaged or obliterated. Consequently, all over
the country, teams of voluntary fire-fighters were set up to help
in emergencies. These volunteers were risking their lives. In
universities staff took it in turns to sit and 'fire watch' in case
the building was bombed or set alight. Blackout instructions
were rigorously adhered to and special permission had to be
obtained before work in the laboratories could be continued
after dark.

Holmes took his turn fire watching in the Durham Science
Laboratories along with the other members of his staff, and
during the day he lectured large batches of RAF cadets who now
came through the department every six months as part of their
course. Unfortunately, unlike during the First World War when
the call to arms reduced student numbers and increased

research time, the need to 'cram' these RAF cadets meant that the long vacations disappeared and precious research time was almost non-existent.

Expected to cover a full year's syllabus in six months, Holmes soon found himself under pressure and looking for ways of speeding up his lectures. The obvious solution was for the students to arrive more prepared having studied some of the work beforehand, but his lectures had always been very self-contained, providing the students with most of what they needed without recourse to extraneous material – the main reason for this being that there simply was no book available that covered all the topics on his course. The three stalwarts of the first year reading list had all been written in the previous century! None of them contained anything about radioactive dating, continental drift or any of the more recent developments in geology. Only now recognising how serious the problem was, it eventually dawned on Holmes that actually there was a book he could use – but it was still in the form of his lecture notes. Immediately he resolved to remedy the situation and to utilise the peace and quiet of fire watching duty to put a book together. Night after night he sat at his upright typewriter – he still did all his own typing even when he had access to a secretary – turning his notes into one of the most celebrated books on geology ever written.

Within a year most of the work was complete and only a couple of chapters remained outstanding. Lack of evidence to the contrary meant that the Earth was by then widely accepted as being, in round numbers, two thousand million years old, so the only other truly contentious issue in geology at that time was continental drift. Holmes thought long and hard about whether to include it in the book for, even in 1943, his theories on convection currents in the mantle as a mechanism for driving continental plates around the globe were still largely ignored, and for a decade now any discussion on continental drift had lain virtually dormant. To come out so publicly in print would leave him exposed to ridicule in a few years time if his theories were proved to be wrong, and a book is much harder to sweep under the carpet than a single paper. He decided to write to his friend Reginald Daly in America.

Although a much older man, Daly, like Holmes, had a background in physics. The two men were also both passionate about igneous rocks, Daly recalling that his interest in geology stemmed from the moment Professor A. P. Coleman held up a piece of granite and remarked, 'This is made of crystals'. Daly had long been an admirer of Holmes and had organised his lecture tour of the States in 1932, whereupon the two men had become close friends and since then had regularly corresponded. It also transpired that Daly too had lost a child at the age of three, his only child as it turned out. So, as he was one of the few Americans who had openly converted to the theory of continental drift, it was natural that Holmes should write to him expressing a moment of doubt:

I have been hesitating whether or not to put continental drift into the book. On the whole, I think not. I am still doubtful about it . . . We really need a first class palaeontologist to assess the biological evidence at its true value. If you could stimulate someone to do that and then take up the problem in the sane light of your long experience we might expect some real progress. However, while it may be complimentary, it is hardly kind to thrust upon you all the headaches and nightmares that such an enterprise would entail!

Daly did not take up the challenge.

In the end Holmes decided that he could hardly ignore what he had been preaching for so long and what, in his heart of hearts, he really considered to be the truth. He had been teaching his students about continental drift for nearly twenty years. Consequently, when they went to conferences at other universities where the old ideas of land bridges and ismuths (island 'stepping stones') were still advocated, they were amazed to discover that what they understood to be the accepted doctrine turned out to be considered heretical and revolutionary.

Furthermore, his ideas had evolved considerably from their formative stage in the 1920s, and it was all beginning to make a lot more sense. In the end he resolved to face the inevitable hostility, not something he had ever shirked before, and the very last chapter of the book summarised all his most recent ideas on continental drift.

While approving of his audacity Daly predicted that 'for this boldness he will doubtless be chastised'. Indeed, he was right. One reviewer, having highly praised the early part of the book, felt that the later section:

is more suitable for rather more advanced students who have developed a capacity to read critically and who are able to distinguish between attractive theory and fact . . . a beginner may be led away by Professor Holmes's clear and attractive presentation and fail to notice that he does not claim that all the opinions expressed in this part are of the same proved or accepted value as the old-established views given in the earlier parts.

Presumably the 'attractive theory' was continental drift, which we now know to be fact, and the 'accepted value' was a 5000 mile land bridge, which we now know to be complete fantasy!

Principles of Physical Geology was published in 1944 and immediately became an international best-seller, despite its rather high price of thirty shillings. The cumbersome title, chosen as a tribute to Lyell and his *Principles of Geology*, was soon dropped by those who bought the book, and ever since it has been known more simply and fondly as 'Holmes'. The first print run of three thousand copies sold out almost immediately and, despite a paper shortage both during and after the war, it was reprinted no fewer than eighteen times in twenty years, becoming the geological bible for generations of geologists. Shortly before his death Holmes revealed the secret of his phenomenal writing success in a letter to a friend: *'Mental effort on the part of the average reader should be reduced to a minimum. To be widely read in English-speaking countries think of the most stupid student you have ever had then think how you would explain the subject to him. It is well worth the additional effort.'*

The extraordinary popularity of the book was due not only to an enthusiasm for geology engendered by Holmes' text, but also to the outstanding quality of the illustrations, largely photographs, which Holmes went to great lengths to obtain, despite many firms having lost all their photographic stock in the bombing raids. The tremendous impact the book made on countless students and amateur geologists is impossible to assess. Undoubtedly some became professional geologists as a result of reading it, as did John Hepworth who enthusiastically summarises how many must have felt:

The very mention of his name [Holmes] is to ignite a spark in my recollections. I first came across The Book (!) in a wonderful parcel of books supplied to the newly established 'Education Initiative' in the Army, at the end of the war (in Germany). I was transferred from the Royal Signals … to the Army Education Corps and among my duties was to receive and unpack this parcel (about 20 books), which was to constitute the regimental library. Among the books was this red cloth-bound volume, with its excellent photos and diagrams, and riveting text – a flash of brilliant illumination in the intellectual murk which we had become so used to in 6 years of 'the Army'.

I had for a long time been interested in geology in a school-boyish way – had inherited my father's Lyell – but this was perhaps one of the most decisive influences in my determining to 'become a geologist'. I am still impressed by the clarity and far-sightedness of his vision.

I recall that as late as the time I graduated (1950) he was still regarded by the establishment as some sort of a maverick – not quite sound – to be spoken of with a smile on the faces of the orthodox. He was not, I think, a Geol. Soc. Man.

Yes, Arthur Holmes – and The Book – must have inspired legions of people.

Indeed they did. Holmes the man and 'Holmes' the book did much to revive failing interest in the geological sciences which, for more than half a century, had been in a rather moribund state. This long period of geological malaise is largely attributable to the fact that the technology needed to solve the big geological problems lagged far behind the theory, so geologists in the early part of the twentieth century had become tired of endlessly discussing questions they were unable to solve. Holmes' problems with the chemical method of analysing lead and uranium were a typical example. Theoretically he knew *how* to date rocks back in the 1910s, but he could not actually do it successfully because it took another thirty years to develop the mass spectrometer. Consequently, as we have seen, Holmes pursued the subject almost single-handedly because few others were interested. No, the things that concerned the geologists of Britain in the early half of the twentieth century were, for example, 'where the Wenlock fitted in' and 'whether the Ludlow should be at the beginning of the Old Red Sandstone'. These details of stratigraphy were the big controversies of the time, because they were questions that could be answered, given sufficient time in the field looking at rocks. And it made little difference to stratigraphers whether the Earth was two thousand or three thousand million years old. But 'Holmes' posed big and exciting geological questions – How did mountains form? What caused earthquakes? Did stable continents wander the globe? – and the Second World War stimulated a revolution in the application and availability of sophisticated instruments in both basic and applied research which enabled some of these questions to be addressed.

Although Holmes' immediate superiors at Durham University recognised the excellent work he had done in building up the geology department out of nothing and so perhaps chose to forget, if not forgive, his private misdemeanours – his affair with Doris – he still encountered hostility from the Council that administrated university affairs. Also, gossip endured amongst the students as each new batch was given the low-down on the private life of the professor and his wife, no doubt greatly exaggerated and embellished with each new telling. But during the war it was difficult to move on. Posts were held open for conscripted men so there was little mobility and no suitable vacancies arose until, in October 1942, the Regius Professor in Geology at Edinburgh University was unexpectedly retired.

Alongside Cambridge University and Imperial College, Edinburgh had once been amongst the top three places in Britain to study geology. But despite its wealth and facilities, Professor Jehu, rumoured to have been given the Edinburgh post because he was a friend of Lloyd George, and his predecessors had allowed its prominence to fade until it had become quite second rate. Thus the university's principal, Sir Thomas Holland, himself a geologist and a supporter of 'continental drift', was determined to use the unforeseen opportunity of Jehu's hurried departure to 'restore the prestige of the Chair to the high level it acquired before 1914 by the work of the two distinguished Geikie brothers', both of whom had opposed Kelvin and his short duration for the age of the Earth. It was a formidable challenge for anyone willing to take it on, and in that context Holmes would be singularly appropriate.

The paucity of vacant posts, especially one as sought after as this, meant that the field was crowded, with twenty-three applications. These highly prestigious 'Royal' professorships, founded by Royal subsidy at a very few British universities, were far and few between and any new incumbent would have to be approved by no less a person than the King himself. Earlier that year Holmes had been elected a Fellow of the Royal Society – only those at the very top of their profession make it into that exclusive club – and so Sir Thomas Holland was quite clear in his own mind which of the candidates he preferred. Going through the motions of making a choice (for he had already written to Holmes encouraging him to apply when Holmes had coyly suggested that fifty-two was too old to start afresh), Holland wrote to William Watts, the now retired Professor from Imperial College, asking his opinion about the relative merits of the twenty-three candidates. By the second paragraph of Holland's letter the initial short list of eleven had been whittled down to five on account of them being the preferred age, but by the third paragraph only one remained:

I am at present favourably disposed to recommend No. 10, who, according to his Vice Chancellor, built up a remarkably good department out of nothing . . . You know the rest of his history, [Holmes had been Watts' student at Imperial College] *which I think justifies the hope that he has the right historical spirit to utilise the atmosphere of the University in which Playfair hatched the Huttonian egg.* [Playfair had been a great friend and supporter of the 'father of modern geology', and did much to publicise Hutton's work after his death.]

It was therefore gratifying for Holland that Watts also thought Holmes way ahead of the field and 'capable of bringing to success whatever he turns his hand to', as did Reginald Daly, by now America's most eminent geologist, who ranked Holmes 'among the best half-dozen teachers of geology within the English-speaking world'. In those days it was important that the professor was a good lecturer because it was always his responsibility to teach the first year class and thus inspire students at an early age, unlike now when they are lucky to see their professor at all in the first year. But in addition to wanting Holmes, Holland also held Doris Reynolds, 'your bright fellow worker', in high esteem. Subsequently Holmes was known to remark that Holland was getting *'two for the price of one'*, for Doris was only ever given an honorary post that allowed her to continue her research, her 'hobby' as some patronisingly called it, but which really meant she worked just as hard as any other lecturer for no pay. Never mind, on £1400 a year, Holmes would be earning more than both of them put together had at Durham.

On the 30th of March, 1943, Holmes received a letter from the Scottish Office confirming 'that the King has been pleased to approve your appointment'. Looking back over his career and all the problems he had encountered over the years – malaria in Mozambique, hostility to his early work, the death of his son in

Burma, poverty and unemployment and the struggle to develop the Durham department – Holmes found it difficult to believe that now, at last, he had finally made such a success of his life. Although not particularly a royalist, he could hardly fail to be flattered at being appointed by the King. Holmes took up the appointment with gusto and by October he and Doris were fully ensconced in Edinburgh.

In keeping with tradition, shortly after Holmes' arrival at the department in Edinburgh, notices were sent out inviting the public to a lecture to be given by the new professor. It was an auspicious occasion; three hundred city dignitaries and prominent members of the University attended Holmes' lecture entitled 'The Age of the Earth'.

In his opening words Holmes paid homage to James Hutton, Edinburgh's greatest geological son whose outstanding achievement, Holmes considered, was his realisation of the Earth's high antiquity. It seemed fitting that Holmes was now walking those same streets, tramping those same hills, and able to observe those same rocks that had caused Hutton to come to the conclusions Holmes could now confirm – the great age of the Earth.

Figuratively *'turning over some of the rocky pages in which the history of our islands is written'*, Holmes described Britain to his audience as it had been in the geological past, moving seas and continents around the globe as easily as computer graphics do today. Never looking directly at his audience, but just over their heads, he gave them the impression he was actually seeing the ancient views he described: icy blue glaciers creeping over the land; seas ringed by golden shorelines, advancing and retreating. Volcanoes puncturing the scene like bursting corpuscles; bubbling flows of lava flooding the countryside. Jungles shrieking with primitive wildlife; seas infested with monsters, the air filled with flying dragons. Deserts appeared and disappeared creating wind-blown sand dunes and ephemeral lakes that evaporated into beds of salt. In tropical swamps the decaying vegetation of countless forests slowly turned into coal. Grey limestones became grey cemeteries for entombed shells and corals, the land once more submerged beneath a warm temperate sea; again retreating, again advancing, until earthquakes shattered the rocks as a vast ocean closed; continents collided pushing up the sea bed into massive mountains in Snowdonia. It was a dramatic picture.

Holmes told the story of Kelvin and his battle with the geologists. He derided the hour-glass methods, showing how present erosion rates are much faster now than in the past. He blamed this on Man, for his poor management of our *'greatest economic asset'* the soil, and for the burning of coal and oil, which *'has for many years steadily increased the potency of the atmosphere to rot away the rocks'*. Finally, he arrived at radioactivity – the new and elegant way of measuring geological time.

Describing how radioactive minerals had become the geologist's timekeeper, natural clocks that kept a material record of the passing years, he told how the oldest minerals so far dated were one thousand seven hundred and fifty million years old, and how meteorites had been accumulating helium for seven *thousand* million years. Time was deep, time was gigantic, time was eternal. But what, exactly, did all it mean for the age of the Earth?

The exploration of the past is still incomplete, for there still remain wide-spread areas of ancient rocks of undetermined ages. At any moment the present age record may be surpassed by new discoveries. All we can be sure of is that our planet has already existed for well over 1750 million years, and probably for not less than 2000 million years – a million times longer, that is, than the whole of the Christian Era. The birth-time of the earth has receded into an inconceivable remoteness and, like Hutton a century and a half ago, we still see 'no vestige of a beginning'.

The audience, many of whom were lay people with no geological experience, was captivated and enthralled. Most had never before stopped to think how old the Earth might be, they had difficulty in imagining time that was so vast, events that were so remote, a world without Man, without beginning or end. Questions were asked and much banter exchanged. Finally, the Professor of Botany stood up to say that at least to him 'it was highly satisfactory that the Earth was older than the Universe!'. Everyone laughed and the proceedings broke up on a note of jollity.

The Earth was older than the Universe?

Astronomers had often been asked to assist in estimating the age of the Earth, and had frequently contributed to the debate, while at the same time trying to calculate for themselves the age of the Universe and various other cosmic bodies such as the sun. During these deliberations, one of the long-standing questions had become: why does the sun shine? What kept it going, giving off so much heat? At the end of the previous century Kelvin had estimated that contraction of the sun due to gravity would keep it radiating (shining) for only twenty million years, and of course he had used this calculation as strong support for his arguments regarding a 20-million-year-old Earth. However, once radioactivity had been discovered and the Earth was shown to be much older than Kelvin had predicted, then the question as to what kept the sun shining became even more critical, for either the Earth must have existed for longer than other bodies in the solar system, or the sun also had to be much older than 20 million years. It therefore seemed logical to conclude that if radioactive elements were responsible for keeping the Earth hot then this was also what kept the sun hot. But in the 1920s when

James Jeans estimated that the sun was radiating mass away at the rate of four million tons per second, it soon became apparent that the energy required to keep the sun shining had to be created by some other, more dramatic process than that generated by the slow decay of radioactivity.

It was not until the 1930s that the wild miracle of nuclear fusion was sufficiently well understood to be considered a satisfactory explanation. At the same time, an understanding of the Universe was also improving. Until the 1920s, astronomers had thought that the Universe consisted of only one galaxy, the Milky Way, and that it was essentially eternal and unchanging. But following Einstein's development of his general theory of relativity it became apparent that the Universe could not be standing still and must either be contracting or expanding. Eventually, the work of Edwin Hubble showed that the Universe was in fact expanding and that the Milky Way, the galaxy to which our solar system belongs, is just one of millions in the cosmos. The obvious deduction from this was that galaxies were moving away from each other as the space between them stretched.

One of the most important implications of these new discoveries was that if the Universe was expanding then, if time was wound back to zero, the Universe must have had a definite beginning; a point in time which could be assigned an age. So, during the 1930s the concept of the 'Big Bang' theory for the start of the Universe was initiated by Georges Lemaître. Unlike the current 'Big Bang' theory which proposes that all the mass in the Universe originates from a singularity no bigger than a pin head, Lemaître proposed that the entire contents of the Universe were packed into a sphere about thirty times bigger than the sun, which came to be known as 'the cosmic egg'. By using Hubble's estimation of the rate at which the Universe was expanding (the Hubble constant) it became possible to calculate the age of the cosmic egg, and hence the time at which the Universe began. Initially, this was determined to be only

1240 million years, already younger than the age of the Earth, then thought to be at least 1600 million years old, but by allowing for errors on both sides, it was generally considered that they were of much the same age.

However, as the age of the Earth gradually crept up towards 3000 million and beyond during the 1940s, this difficulty of an Earth older than the Universe became acute. The problem lay of course in accurately determining the Hubble constant, but such confidence was placed in Hubble's data that for years no-one seriously considered exploring the possibility that it might be wrong. In 1936 Hubble himself concluded that 'further revision [of the constant] is expected to be of minor importance', and as late as 1949 it was considered 'highly improbable' that observational changes in the value of the Hubble constant would lead to a resolution of the time scale problem. Indeed, the age of the Universe was only increased to 1800 million years, by which time the age of the Earth was almost twice that. The only solution seriously considered at the time by some astronomers was to resurrect John Joly's ideas that the law of radioactive decay had changed with time. If radiometric dates could not be trusted then the age of the Earth was probably much younger than that estimated by geologists – and anyway everyone knew that geologists could not count! For years, Holmes and his consistent dating of the Earth continued to be a thorn in the astronomers' side.

The problem lasted well into the 1950s, when new measurements of the Hubble constant finally extended the age of the Universe to a point where it was safely older than the age of the Earth, and for many years after that it was considered to lie in the 10–15 billion year range. But in 1994 measurements made from the Hubble space telescope suggested an age as low as 7 billion years for the Universe, causing momentary interest by the world's press, who questioned the validity of the Big Bang theory. Today, we accept that the Hubble constant is a very difficult value to measure and that no doubt it will continue to fluctuate for many years to come; the only thing we can say with any confidence is that the age of the Universe is unlikely to ever again be younger than the age of the Earth.

While astronomers may have been disquieted during this episode, the geologists were unperturbed and confident in their own radiometric dates, which were now widely accepted as being reliable. The problem of a young Universe was not theirs and they even delighted that this branch of physics was unable to get its sums right, when so often the geologist had been criticised by the physicist for his own numerical ineptitude!

The Age of Uranium

Because the pathway from uranium to lead was peculiarly
complicated, others had abandoned their researches, leaving
the 21 year old research student to become the world
authority on a technique that was finally to provide the
planet with its authentic, scientifically determined birthday.

Robert Muir Wood on Arthur Holmes

Half a lifetime had gone by since Arthur Holmes had lain in his
tent in Mozambique, racked with fever, dreaming of developing
a geological time scale and wondering how he could reconcile
the age of the Earth as determined by radioactivity with that
calculated by the old established methods of sedimentation rates.
While progress on a geological time scale had been made over
the following years, it had largely been in the physics arena:
improved understanding about the atom; the discovery of isotopes; development of the mass spectrometer; and recognition
of the four stable isotopes of lead. The geological side, however,
lagged far behind. A rock assigned an age of 300 million years,
for example, still could not be classified as 'Carboniferous' with
any confidence because it was still not known how long, in
geological time, the Carboniferous ranged. So, as we saw with
the helium results from the Whin Sill, extreme errors could be
accepted as reasonable values because no limits could be placed
on the extent of the Carboniferous. Clearly, what was needed was
a time scale that said 'the Carboniferous starts here at this age
and ends there at that age, therefore any age in between must be
Carboniferous'. But that was still a long way off.

In the case of the Whin Sill it was the geological age that was correct and the radiometric date that was seriously in error because of the dating system used. However, at the time there was no way of knowing this so William Urry wasted five precious research years working on a method that was never going to produce accurate results. But science progresses like that – it is often a gradual process of elimination that eventually reveals the truth after many years of assiduous hard work. Unfortunately, it is sometimes easy to get discouraged along the way when results are not forthcoming, and by 1938, after the failure of Urry's five-year effort to develop a scale based on his extremely disappointing helium results, no university anywhere in the world retained a programme designed to build a geological time scale.

The difficulty was still due to the same things – ages could still only be obtained from igneous rocks that contained unusually high amounts of lead, in itself a problem due to the scarcity of these rocks, and when rocks suitable for radiometric dating were found, poor geological control meant uncertainty as to what geological age they were. The nature of igneous rocks is that they tend to cut across or intrude into the very sedimentary rocks that contain the fossils used for determining the geological age of the igneous rocks that cut them. The arguments become somewhat circular. As a result very often all that could be said of the geological age of an igneous rock that intruded a sedimentary rock was that the igneous rock must be the younger.

But there was light at the end of the tunnel, and once again it was a physicist who was holding the torch. With the recognition by Rutherford back in 1929 that there were two isotopes of uranium that each decayed to a different lead isotope it was soon realised that the relationship between uranium and lead contained not one, but two geological clocks that could be used to verify each other – the decay of uranium 238 to lead 206, and the decay of uranium 235 to lead 207. Not only that, but because

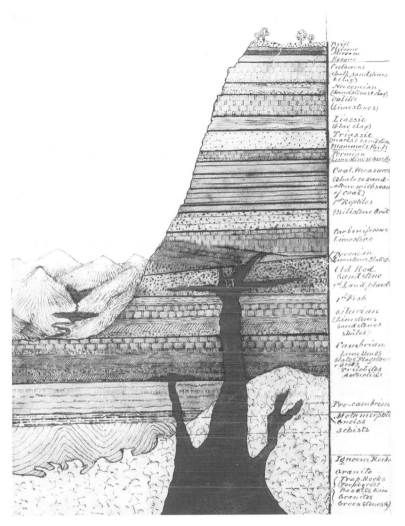

The Dhustone section.

This geological section illustrates the difficulties of assigning a
geological age to igneous rocks. In this case, all that can be said of
their age is that they must be younger than Devonian, the last lay-
ered sedimentary rocks they intrude and cut across. They could of
course be as young as Tertiary, but not have penetrated any further
than the Devonian. The originator of this section is unknown, but
it is believed to be from mid-Victorian times and to represent the
famous Shropshire dhustone rock near Ludlow.

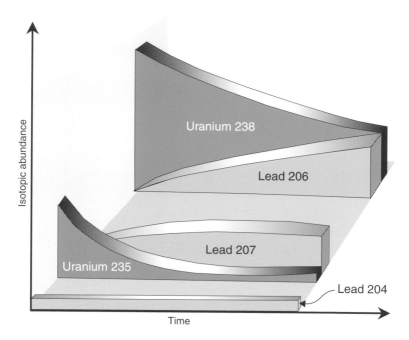

Nier's three geological clocks.
Clock 1, uranium 238 → lead 206. Clock 2, uranium 235 → lead 207.
Clock 3, the growth of lead 206 and 207, relative to 204.

the two clocks were ticking at different rates (the decay rate of uranium 235 is more than six times faster than that of 238) it was possible to determine the age of a rock simply by comparing the growth over time of the two resulting lead isotopes, relative to something else that did not change with time. So when Alfred Nier finally identified 'ordinary' lead (lead 204) it became that 'something else' because it was not derived from radioactive decay and therefore did not change in value over time. In other words a third clock was found simply by measuring the amount of lead left by each of the two uranium isotopes and then comparing them with 'ordinary' lead to see how much they had increased over time. This new 'lead–lead' method, as it was called, was a major step forward and is still one of the fundamental age dating techniques used in modern geochronology.

To demonstrate the validity of using these three geological clocks now available to him, in the late 1930s and early 1940s Alfred Nier performed a series of very precise age determinations on twenty-five different rocks of varying geological ages, from different parts of the world, using the most up to date mass spectrometer then available to him. While in most cases the three ages for each 'clock' agreed within reason, one rock, the Manitoba pegmatite from the Huron Mine in Canada, gave a wide range of age values, the most reliable of which was thought to be the oldest one at 2200 million years. (A pegmatite is a coarse grained igneous rock that frequently contains the world's largest crystals and choicest mineral specimens.)

The implications of finding this extremely old mineral meant that once again a large question mark hung over the age of the Earth, which, by then, was considered to be fairly well established at 2000 million years. Such was Nier's concern about the age of this mineral at a time when astronomers were insistent that the Universe could not be more than 2000 million years old (two billion), that a year later he analysed another rock from the same area in an attempt perhaps to *disprove* his first age. But instead he obtained an even older age – this rock was 2570 million years old! Even more worried, Nier was reluctant to trumpet his results and modestly claimed that 'One of the samples was the oldest so far studied and appears to have an age close to two billion years.' Actually, if he were being strictly scientific, an age of 2570 million years is closer to three billion than two billion years. Unfortunately, intervention of the Second World War prevented Nier from investigating this problem any further when he was required to assist in development of the atom bomb.

The nuclear fission of uranium had been discovered in 1939 by Enrico Fermi, a Nobel-Prize-winning Italian physicist then working in America to escape the rise of Fascism. Nuclear fission is the splitting of the atomic nucleus, which is accompanied by the emission of two or three neutrons and the release of large amounts of nuclear energy. While the process does occur spontaneously in nature, it can also be induced by bombarding nuclei with neutrons. The neutrons released in this process are then used to induce further fissions, setting up a chain reaction that must be controlled if it is not to result in a nuclear explosion.

The potential for nuclear fission to be used as a weapon during the war was obvious, but initially there was some question as to which of the uranium isotopes was responsible for fission. So Fermi urged Nier to try and separate the uranium isotopes using his mass spectrometer, and help determine which one could account for the slow neutron fission he had observed, stating that to know which was which was 'of considerable theoretical and possible practical interest'. When the analysis was duly performed (using a geological sample originally designated for age determinations as the source of the uranium) and uranium 235 was shown to be the culprit, Fermi went off to build the first nuclear reactor. Following that, in 1941, when it became apparent that enriched uranium 235 might be used for making atom bombs, Nier's laboratory contained the only mass spectrometer in the world that could analyse uranium isotopes. Inevitably, he was required to perform analyses for various groups working on such problems and so found no time for geologically related research. But it was during this period that Holmes renewed their correspondence.

Intrigued by Nier's results from the Manitoba pegmatite, Holmes had re-calculated Nier's age data for himself and found the pegmatite to be 2480 million years old, a value closer to the older result determined by Nier. He wrote to tell Nier in May 1945, and cautiously expressed his views:

I hope when you get going again [after the war] *you will try to clear up the extraordinary discrepancies in the Manitoba results. I feel sure that the true age must be of the order of 2000 m.y. or more, and this is of the greatest interest, not only because the rocks here seem to be the oldest yet found, but also because such a figure shows that current views about the expanding universe need revision – not perhaps to be wondered at!.*

Indeed, while astronomers may have been prepared to stretch a point and agree with the geologists that 'within error' the age of the Earth and the Universe were more or less the same at 2000 million years, Holmes did not feel constrained by the astronomers to do likewise. If the age of the Manitoba pegmatite meant that the age of the Earth became at least 500 million years older than the Universe, then it must be the astronomers who had got it wrong. For what was evident to Holmes as a geologist but not to the astronomers nor to Nier as a physicist, was that the Manitoba pegmatite was the *youngest* rock in a much older sequence.

Geological investigations had shown that the sequence had started to accumulate in a primordial ocean when a huge thickness of sediments, some several kilometres deep, was deposited on the ocean floor as an ancient continent was being eroded away. These sediments had been buried to great depths where they became very hot, partially molten, and metamorphosed into a gneiss. Uplifted during a period of mountain building, the gneiss was then intruded by multiple phases of molten lavas until finally the residual fluids and gases, now highly concentrated in elements such as uranium and lead, crystallised out to form the Manitoba pegmatite. Holmes realised that the whole process from deposition on the ocean floor to final formation of the pegmatite must have taken many hundreds of millions of years, so if the pegmatite itself was giving an age around 2500 million years, then the gneiss it intruded, which had once been ocean floor sediments, may well be 3000 million years old.

And what about the continents the sediments had been derived from? They must have been even older! It was an exciting prospect.

Holmes pondered these thoughts. He had been extremely impressed with the quality of Nier's work on the Manitoba pegmatite and the other twenty-four samples, and entertained the hope that from these precise data *'it might be possible to fathom the depths of geological time'*. He recognised that the twenty-five samples not only gave information about their individual ages, but that locked up in the data could be evidence pertaining to the initial state of the Earth, its 'primeval' state, if only he could extract it. It was time to resurrect an old idea.

Fifteen years earlier, in 1932, when working on the possibility of using another decay scheme, the decay of potassium 41 to calcium 41, as a new isotopic system for dating rocks, Holmes had devised the principle of using primeval (or initial) isotope ratios to calculate *'the time when the separation of the granitic and basaltic layers took place in the newly-formed earth'*. In essence the idea was to determine the original ratio of two isotopes in any isotopic system as it was when the Earth first formed. By knowing the rate of decay of that system, the time taken for the initial ratio to evolve to its present day value would essentially be the true age of the Earth.

Similarly, it would be possible to determine, for example, whether a granite had been derived from a basalt or from melting of the crust, simply by knowing the isotope ratio of the granite today, and back-calculating to see what the initial ratio had been when the melt first formed. A granite derived from melting the crust would contain far more potassium than one derived by differentiation from a basalt, so the ratios of potassium to calcium isotopes, after a significant period of time had elapsed, would be demonstrably different and could be used to indicate the source material. Holmes recognised that the idea had tremendous potential not just for dating the age of the Earth, but for solving many problems related to the origin of igneous rocks and termed it *'A New Key to Petrogenesis'*.

At that time, however, the isotopes of potassium had only just been identified and it was still not certain which ones were radioactive, so Holmes was unaware that potassium 41 is in fact the stable isotope of the potassium family and therefore cannot decay. Once this flaw was recognised, the paper and its new key to petrogenesis became redundant and lay ignored by the geological community, although the principles of the method were still perfectly valid. It was not until the 1960s, when the idea was 're-invented', that 'initial ratios' became one of the foundation stones of isotope geochemistry. True to form Holmes has never been given the credit for it, although the idea is possibly one of the most remarkable examples of perspicacity in geological research. Perhaps embarrassed by his error, he himself allowed the model to lie dormant for fifteen years, but when attempting to extract the age of the Earth from Nier's data he resurrected the concept again.

In principle the problem he was hoping to address was fairly straightforward: from the moment uranium and thorium were formed somewhere out there in the cosmos, radiogenic isotopes of lead would be added to the 'ordinary' lead that had formed at the same time as uranium, giving rise to a mixed composition Holmes called 'primeval lead'. The isotopic ratio of this primeval lead would then continue to evolve as long as uranium and thorium continued to decay, until the primeval lead became separated from its uranium and thorium source, probably during formation of the Earth's crust. At this point the composition of the primeval lead would become 'fossilised' or frozen for ever more, held inside minerals within rocks forming the Earth's crust. In theory therefore, it should be possible to find minerals within ancient parts of the crust that still contained this primeval lead composition. Having identified the composition of primeval lead, and knowing the rate at which uranium added radiogenic lead to the system (the decay rate of uranium), it should then be possible to calculate the time that had elapsed between the 'fixing' of primeval lead in the crust and its present day values – which would essentially give the age of the Earth.

Unfortunately, Nier considered that the majority of his twenty-five samples contained a mixture of both primeval lead and radiogenic lead, the latter having contaminated the fossilised primeval value as uranium and thorium continued to decay within the Earth's crust. But there was one sample, a very ancient galena (lead ore) from Invigtut in Greenland (where some of the oldest rocks in the world are now found), that contained very low lead ratios and no uranium or thorium. These factors suggested that it might just be a relict of primeval lead that had become trapped in the Greenland rocks at the same time as formation of the Earth. So, making the assumption that the Invigtut galena did indeed represent primeval lead, and using Nier's other twenty-four samples to represent the isotopic constitution of lead as is was at various ages throughout Earth history, Holmes once again set out to date the age of the Earth.

The calculations were complex and immensely time-consuming to do by hand, so he applied to the University for a grant with which to purchase a Marchant calculating machine. A whole year passed before it was sitting on his desk. Silent and speedy, it was £74 8/- well spent!

On the 16th of February 1946 Holmes wrote again to Nier to tell him some exciting news:

Ever since your isotopic analyses of ore-leads was published I have hoped that it would be possible to calculate from the results the time that has elapsed since the Earth's primeval lead began to be contaminated by radiogenic lead. The acquisition of a calculating machine a few months ago has now made possible the somewhat formidable calculations and I have just completed the work. The age works out at about 3,000 million years by various sets of solutions . . . the average of the best set of solutions being 3015. We can however, afford to neglect the odd 15! This looks like being the first

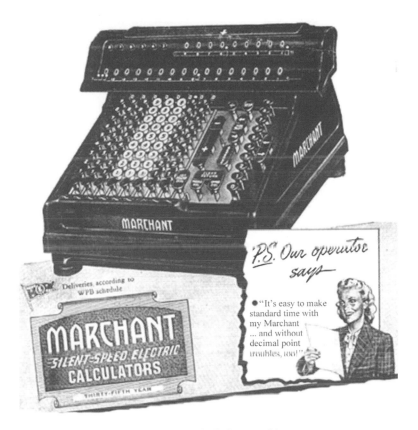

A Marchant Calculating Machine.
The size of a foot-stool and 'representing the apex of achievement
in 36 years of continuous improvement in the calculating machine
art', this electrically operated and fully automatic calculator
performed multiplication, division, addition and direct subtraction
at 1300 counts per minute, 'with all keyboard controls conveniently
grouped under the fingertips of one hand'.

*really reliable estimate of the age of the Earth and I should
like to salute your work as the means of making it possible.*

This letter marks a defining moment in the history of dating
the age of the Earth. A wild miracle had been performed not,
like many of the others, in a blinding flash of revelation, but as

the result of dedicated hard work over many years of considerable trial and much error. The combined talents of these two great scientists – the fastidious care with which Nier performed his analyses, and, with good data to hand, Holmes' ability to think deeply about the great geological problems of his day – eventually provided us with a means whereby the age of Mother Earth could at last be deduced. As yet, they had not got that age right, but everything was in place – the basic technology to obtain the data and a model to calculate the results. All that remained now was to improve the machine and find the right samples to work on.

But as usual, not everyone liked the idea of an ever more ancient planet. It seemed to some that just as they were getting used to the idea of the Earth being a certain age, then up it would go again, requiring them to once more readjust all their preconceived ideas. A typical reaction was as follows:

We must congratulate him [Holmes] on having again advanced our knowledge of geological ages by a marked achievement. But I believe it should be emphasized that what has been ascertained is not, as Holmes states, the 'age of the earth', but the 'age of the materials forming the earth'. Our conceptions of the birth of the solar system and the earth are vague, but in current opinion the zero time that Holmes has determined would not apply to the earth but to some earlier event, that for convenience might be called the birth of the Milky Way. The subsequent formation of the earth as a separate entity came later and there is no reason for supposing that the clock Holmes has read for us, was set to zero again by that revolution. Hence, the earth is younger than 3,000 million years. The difference need be only a small amount, but it may also amount to a large fraction of the estimated age.

I hope this remark will serve to clarify a minor point in an otherwise lucid and epoch-making contribution from the great time keeper among geologists.

Holmes was slightly indignant: 'I had not overlooked the fundamental distinction to which Professor Kuenen directs attention', and went on to clarify his reasoning. 'It is generally *considered to be highly probable that the earth was originally gaseous and that the period of consolidation, up to the time of formation of a solid crust* [Holmes called this the granitic layer], *was relatively short. Jeffreys, for example, estimates that "the earth probably became solid within 15,000 years of its ejection from the sun". Even if this estimate were wrong by a factor of a thousand, the age of the granitic layer would not be appreciably different from the age of the earth* [in fact it would be equivalent to the 'odd fifteen' he had suggested they neglect when writing to Nier!]. *Accordingly the age I have determined refers to the time when the granitic layer separated from average earth material during the consolidation of the globe ... and that, for all practical purposes, is indistinguishable from the age of the earth.'*

Having thus established the age of the Earth by assuming the Invigtut galena to represent primeval lead, Holmes had then projected his ideas backwards in time to a point at which uranium *'first began to deteriorate'* and when the lead ratios were essentially set to zero. He deduced this *'age of uranium'* to be 4460 million years ago. By this reckoning, if the age of the Earth was 3000 million years, then the 'short' period Holmes mentions between consolidation from the gas cloud and formation of the granitic crust must have been preceded by a very 'long' period of more than a billion years while uranium decayed within the gas cloud, a fact that does not seem to have interested him at all. But, as Kuenen pointed out, concepts regarding the birth of the solar system were 'rather vague', so Holmes did not concern himself unduly with events that occurred before the Earth consolidated. Ten years later he must have kicked himself when he realised just how close he had got to the right answer for the age of the Earth, which he had interpreted as 'the age of uranium'.

A few months after making these initial calculations, with his Marchant calculating machine red hot from a revised estimate based on no fewer than 1419 solutions for *'the time since the isotopic constitution of the Earth's primeval lead began to be modified by addition of lead isotopes generated from uranium and thorium'*, Holmes concluded *'that on the evidence at present available, the most probable age of the Earth is about 3,350 million years'*. It was slowly going up – only another billion or so to go.

Nier's precise radiometric data had triggered a tremendous amount of interest amongst geologists, and other workers in the field recognised the opportunity of taking the 'Age of the Earth' prize for themselves. Fifty years previously the race to date the age of the Earth had fallen by the wayside with the discovery of radioactivity, but now it gained momentum again, if with a smaller field of runners, and Holmes was not the first person to use Nier's data to try and deduce an age of the Earth. In 1942 a Russian, E. K. Gerling, had attempted it and arrived at a minimum value of 3950 million years, but as his work was published in Russian Holmes and others did not become aware of it until it was translated into English, years after they had independently arrived at their own estimates. Similarly, a German, Fiesel Houtermans, published his 'age of uranium', which he found to be 2900 million years, very close to Holmes' value for the age of the Earth. Although Houtermans acknowledged that Holmes had published a couple of months before him, because the techniques they both employed were very similar, the model they adopted soon became known as the Holmes– Houtermans model for dating the age of the Earth.

But, although all three came fairly close to the right answer, they had all made the same fundamental mistake: at the end of the day neither the Invigtut nor any other sample of lead found in the Earth's crust contained the magic primeval lead value. Nevertheless, the mathematical model they developed to do the calculations with was elegant and perfectly valid and is essen-

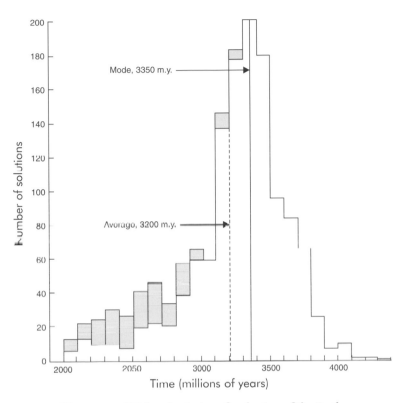

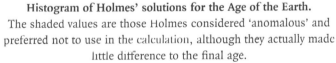

Histogram of Holmes' solutions for the Age of the Earth.
The shaded values are those Holmes considered 'anomalous' and
preferred not to use in the calculation, although they actually made
little difference to the final age.

tially the same model still used today for calculations of the age
of the Earth.

When being presented with one of his many medals for outstanding achievements in geology, Holmes described how his
greatest disappointment, his work on the potassium–calcium
decay system, turned into his greatest satisfaction, that of dating the age of the Earth from Nier's isotope data. The model
used was fundamentally the same on both occasions, only the
isotope system was different. We do not know whether

Houtermans knew of Holmes' paper on 'The Origin of Igneous Rocks', published in the early 1930s, where Holmes first proposed his ideas on initial ratios, or if Houtermans arrived at his ideas independently from Holmes in the late 1940s, but either way there is a strong case for renaming the 'Holmes–Houtermans' model for dating the age of the Earth as simply the 'Holmes' model, since Holmes clearly got there first.

While waiting for the arrival of his calculating machine so that he could complete the formidable calculations required to estimate the age of the Earth from Nier's data, Holmes turned his attention once again to his old dream of building a geological time scale. By using Nier's data as control points in the geological column, he felt that at last the time had come when he should be able to resolve the old conflicts between the hourglass and the radiometric methods *'in a reconciliation of converging evidence'* from the base of the Cambrian up to Recent times. The problem however, was twofold: the majority of Nier's results were on Precambrian samples, in other words they were too old, and of those that were younger than Precambrian, the geological ages were not well constrained. This process of elimination left Holmes with only five control points out of Nier's twenty-five samples to use in the construction of his scale, so how was he to estimate the duration of each geological period?

Samuel Haughton's celebrated principle of 1878 said, 'the *proper measure of geological periods is the maximum thickness of the strata formed during these periods*' but at that time a rigid interpretation of this principle assumed that the rate of accumulation for any particular rock type was always the same, whatever the conditions it formed in, and, as we have seen, this led to wildly inaccurate estimates for the age of the Earth. If however, one accepted the fundamental premise that the maximum thickness of strata formed during any one geological period was indeed some indication of the time taken to form that period, but one also had independent constraints on that time, then it should be possible to combine time and thickness to build a geological time scale.

The problems inherent in such a method were well recognised by Holmes and he pointed out that any estimate of maximum thickness was likely *'to be full of innumerable "gaps" representing intervals of erosion'*, and thus much of the true thickness of the rocks could well be missing. He also acknowledged that there may well come a time when even Nier's precise data were superseded by superior methods. Nevertheless, in 1947 there appeared to be no other way of providing a scale for the duration of geological time, other than by using the isotope dates available as 'way points' in the total thickness of sediments. So, having collected as much accurate data as he could find on sediment thickness from around the world, and having spent a great deal of time evaluating the 'most probable age' represented by each of the five samples selected from Nier's data, he decided the approach worth trying while *'remembering that neither the maximum thicknesses so far reported nor the most probable ages adopted as control-points are likely to be final'.*

It was an incredibly simple idea. Taking a large piece of squared paper he first drew to scale the total thickness of rock known from each geological period, until he had the whole geological column from the present day down to the Base Cambrian represented by the thickness of rock it contained. Against this he then plotted the 'most probable ages' derived from Nier's five data points in their 'most likely' geological positions, and drew a curve from top to bottom through the fixed control points. Suddenly, in a few short minutes, after half a century of hard work, there, lying on the table in front of him was a geological time scale. The whole of his life's work had been directed to this one exhilarating moment. The dream had finally come true. A 'true' age could be obtained for any rock with a known geologi-

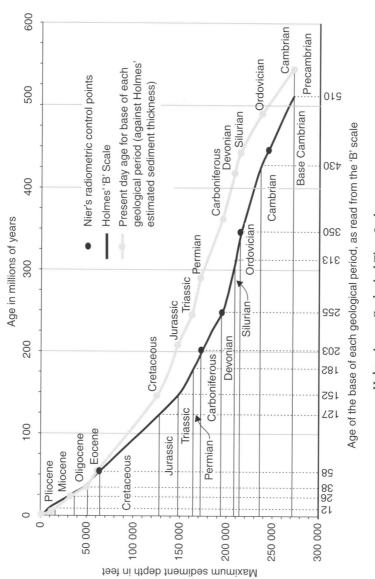

Holmes' 1947 Geological Time Scale.

The lower curve is that deduced by Holmes. The upper curve was derived by plotting present day values for the base of each geological time period against the same 1947 values for maximum sediment thickness as estimated by Holmes.

Holmes' 'B' Scale	Ages in millions of years	Today's values
1	Pleistocene	1.6
12	Pliocene	5.1
26	Miocene	24
38	Oligocene	38
58	Eocene	55
–	Palaeocene	65
127	Cretaceous	146
152	Jurassic	208
182	Triassic	245
203	Permian	290
255	Carboniferous	362
313	Devonian	418
350	Silurian	443
430	Ordovician	490
510	Cambrian	544

Holmes' 'B' scale age determinations.
The table compares the age values determined by Holmes for the
base of each geological period, as read from the 'B' curve, with
present day values.

cal age, and the duration of each geological period could be read
directly from the scale.

It was a brave first attempt but, as Holmes had predicted, the
accuracy of Nier's 'most probable ages' did improve with time
and greater sediment thicknesses were discovered elsewhere.
However, the main cause of discrepancy with today's measurements was the age-old problem of the poor geological control
of his five data points. These turned out to be significantly in
error – a difficulty that somewhat undermined their value!
Nevertheless, the 'B' scale, as Holmes called it, enjoyed an unexpected success for ten years in a wide variety of scientific fields,
setting a standard that no-one else at the time was able to

improve upon. Thirteen years later when Holmes came to revise the scale he was quite happy to acknowledge that it had outlived its usefulness and joked that *'now I come to bury the B scale, not to praise it'.*

Both Arthur and Doris worked extremely hard, trying to build up the geology department in Edinburgh to its former glory, but perhaps inevitably this interesting and intellectually vigorous couple once again attracted tremendous controversy. This time it was about their work. For years Doris had been working on granites, and the question that puzzled her most, and many other workers in the field, was 'how had they originated?'. In Britain we are familiar with granites in places like Dartmoor and Cornwall, where granite tors stand out above the landscape. These are in fact only the surface expression of a much more massive body of granite that probably links all the tors together underground and which may extend many miles down into the crust of the Earth. In other parts of the world, the Himalayas for example, the lateral extent of these massive bodies is exposed at the surface and can be traced for hundreds of miles. They are the roots of ancient mountain chains, now long gone, but in the 1940s little was understood about their origins.

Doris had a theory; she called it 'granitisation'. The model proposed that fluids migrating upwards in the crust from within the bowels of the Earth, 'a flux of emanations' as she called it, soaked into the existing rocks adding new elements and thereby turning the rock into a granite. One of the features of this theory was the 'basic front', which formed as a halo around the altered rock as the incoming fluids drove out dark elements such as iron and magnesium. Norman Bowen, a Canadian geologist and leader of the opposing theory, that granites evolved from basalts, described the basic front as a 'basic *affront*' to much popular acclaim! The two schools of thought became

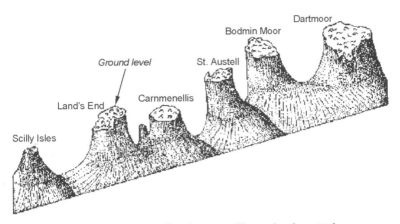

The granites of Cornwall and Devon, illustrating how Holmes considered their roots were connected underground.

polarised as the Edinburgh geology department became more and more deeply 'granitised'. All the Holmes' students 'did a granite' for their PhD thesis and the very mention of an alternate theory meant you were liable to have your head chopped off by the Queen of Granites. It was not a very healthy state of affairs.

At the other end of the country Bowen's followers were concentrated in the Cambridge-dominated Mineralogical Society where the word 'Holmes' became an anathema. At one meeting the situation had become so antagonistic that Doris was forced to buy herself 'one of these modern high hats like a witch. I thought that if I kept that on at the meeting I could not be overlooked.' There were no half measures with Doris, she was an extremely strong and forceful personality, which undoubtedly helped her to succeed in the intensely chauvinist and male-dominated profession that she found herself in, but of course that very success alienated many of her male colleagues. Her students though, seem to have adored her.

Arthur appears to have supported his wife in these theories, continually praising her work both in print and in lectures – *'she has made unique and outstanding discoveries in granitisation'* – although it is not clear how much he really believed her theories. They only wrote one paper on the subject together, and that was published in an obscure Finnish journal; nevertheless, he inevitably became tarred with the same brush. He certainly fell from grace in the eyes of many geologists both for his apparent belief in granitisation and for his adherence to the theory of continental drift, still not accepted by most geologists in the 1940s and 1950s. But Holmes was more than used to having his work criticised; it was like water off a duck's back, and it seems that Doris was of the same breed. Undoubtedly though, even if Arthur did not fully believe in granitisation he would support Doris to the last for her right to be innovative and courageous, and for erecting 'wickets to be bowled at'. Today fluids are considered to play an important role in geological processes within the Earth's crust, and continental drift, embodied in its latest guise as the 'Theory of Plate Tectonics', ranks as one of the greatest unifying theories of all time – but this hindsight was still to come.

More onerous from Holmes' point of view was the fact that the job of Regius Professor of Geology at Edinburgh University was proving far more arduous than he had imagined it would be, and it was beginning to tell on his health. He wrote apologetically to his American friend Reginald Daly:

I should have written to you long ago about various matters of mutual interest, but I expect you can readily forgive me when you remember how overwhelming the work of a Professor can be. These last few years have become worse than ever: each day bringing more work than can be got through and leaving a vast pile of arrears for vacations. Departmental duties use up all my energies and were it not for vacations I should get little 'real' work done.

Daly responded sympathetically:

Yes, the demands on an active Professor in a great university seem nowadays to have no limit. As with you, the best men [at Harvard]

Doris and Arthur Holmes on the Giant's Causeway, Northern Ireland.
They spent every summer doing field work in Ireland, and it was
there that Doris developed her ideas on 'granitisation'.

*have been victimized – with tendency to convert them into glorified clerks
or, more politely expressed, administrators. That nevertheless you continue to turn out fundamental results in research is a fact that keeps me
ever in mind of your great power as a leader in the New Geology!*

In similar vein, Nier and Holmes frequently apologised to
each other, Nier often not replying to Holmes for up to a
year.

Arthur's health had never been very good since Mozambique,
and undoubtedly it was not improved by the winter of 1947, one
of the severest on record. The snow was so deep that people
walked along the tops of the hedges in areas where it could not
be cleared. Then temperatures plummeted to minus 16 degrees
centigrade and the country stayed frozen until March. Most
foodstuffs were still being rationed in the aftermath of the war;
there was a transport strike and life became very strenuous as it
became almost impossible to get around. To cap it all the
rationing of coal was increased, so keeping warm was difficult

too. People chopped up old furniture to put on the fire, wooden fences disappeared over night, and trees seen standing one day were no longer there the next. In the Edinburgh geology department a large piece of coal exhibited in the main entrance hall, displaying the beautiful fossil of a three hundred million year old fern, mysteriously disappeared one day. No one complained; they just wished they had thought of it first. So it is perhaps not surprising that during this severe weather Holmes was taken seriously ill with what he described to Nier as a 'breakdown' and to Daly as a 'kind of climacteric.' There was nothing discoverably wrong, but he lost all energy and, worst of all, all interest in his work. After trying to stagger along for a few months he had at last to give in and take a long holiday. The doctors ordered a complete rest so he and Doris spent all summer in Ireland, where he recuperated amongst the rocks of their beloved Donegal. 'Since then I have been reasonably fit, though strictly on condition of living at a slower rate than previously. I can think of lots of things to do, papers to write and so on, but doing and writing them is another matter altogether.' Needless to say, he does not appear to have slowed down in his prodigious output of work but, having extracted as much as he could from Nier's data, he turned his attentions elsewhere.

As far as Holmes was concerned he had found the age of the Earth to be 3350 million years, and even if that date was not exact, he had provided the model that would allow others to get it right once better data were available. Furthermore, he had fulfilled his life-long dream of developing a geological time scale that could be applied to common rocks and, even more significantly perhaps, achieved what had once seemed impossible – reconciliation of the hour-glass methods with radiometric dates. All that remained now was to improve the techniques and refine the ages. That could be left to a younger generation.

Arthur Holmes had worked his wild miracles quietly and persistently over many years of trial and error. Finally he had shown the world how to tell geological time.

The Age of the Earth

If one is sufficiently lavish with time, everything
possible happens.

Herodotus

Back in 1913 Arthur Holmes, then a young man of twenty-three, had just published his first book on *The Age of the Earth*. While writing it he had come across a theory with regard to formation of the Earth recently put forward by an American geologist, Thomas Chamberlin, who considered that the Earth had been created by the accumulation of cold solid particles which Chamberlin called 'planetesimals'. In Holmes' mind the most important feature of the Planetesimal Hypothesis was that Chamberlin rejected the assumption shared by Kelvin and other scientists that the Earth had begun as a molten globe. Instead, Chamberlin proposed that although heat would initially be generated by planetesimals falling into the Earth as it consolidated, during the later stages that heat would be dissipated into space leaving the Earth a cold and solid body. The particular attraction of this theory for Holmes was that it discredited Kelvin's arguments in favour of a cooling globe and a very restricted geological time scale.

Thrilled that someone else shared his views of an ancient Earth, Holmes wrote to Chamberlin in 1912 to tell him *'how much your work on cosmogenic geology and causal processes has inspired me in my geological work. You have done more probably than any other living geologist to clothe the dry bones of geological fact with the fascination of co-ordinating theories'.* In the same letter he predicted with remarkable foresight for one so young that *'ultimately we may drop the use of sedimentation and salinity etc. as methods of measuring time, in favour of that which employs the rate of radioactive disintegration as its chronometer'.*

Thus inspired, the young man realised that a study of meteorites might help him understand the origin of the Earth and its crust: *'It is manifestly impossible ever to know directly the chemical constitution of the Earth's interior. However, we may study in the laboratory the disrupted fragments of some other world, for it is now believed that meteorites were once . . . parts of a cosmic body. If this be true it seems highly probable that the constitution of the meteoritic parent body thus determinable was essentially similar to that of the earth.'* So he wrote to Dr Prior at the British Museum in the hope of obtaining some meteoritic material to work on: *'I am attempting at present to work out some deductions from the Planetesimal Hypothesis of Chamberlin, concerning the evolution of the Earth's crust. Analyses of meteorites are of great use in this investigation.'*

Unfortunately, samples of meteorites were difficult to get hold of. Dealers would sell them for astronomical prices that were usually well out of the range of university budgets, and the British Museum was reluctant to part with those offered to them. A letter from a Major R. Archer Houblon describes one such offer:

On Thursday 12th last a meteorite fell close to my house. The hole made exactly resembles the whole [sic] made by a "dud" shell. It fell with a blinding flash and simultaneously a very violent detonation, the concussion killing 3 sheep. Before digging it up, I propose leaving it until I hear from you incase you wish any particular report or investigation made.

In the end the specimens used by Holmes *'were detached from a small collection which belongs to the Geological Museum of the Imperial College'*.

Meteorites fall into two main categories – iron and stony – which are further divided into classes. Holmes argued that *'for each type of meteorite there corresponds a terrestrial zone* [on Earth]*'* and set out to illustrate this by making some exquisitely delicate analyses of the tiny amounts of radium in meteorites. Making the assumption that iron meteorites represented part of the core of a once much larger body with a stony exterior, in the same way that the Earth has an iron core with a stony exterior, he compared radium values obtained on meteorites with those from rocks on Earth, and showed that in both cases the amount of radium decreased towards the core. Furthermore, the radium values in the stony meteorites were found to be very similar to those found in rocks known to come from deep within the Earth, and because no radium was seen in the iron meteorites he inferred that none would be found in the Earth's iron core. A genetic link between the Earth and meteorites was clearly indicated by this radium study, and the tremendous significance of this work was not lost on Holmes. He concluded: *'The prob lems that are suggested by . . . this inquiry are of supreme geological importance. The evolution of the Earth, of its zonal structure, and particularly of its crust are all questions which remain to be solved.'* Indeed, most of those questions are still being addressed today even though Holmes was to devote a great deal of time to them over the following years, profoundly contributing to our understanding of the Earth and the evolution of its crust.

But the amount of meteoritic material available to Holmes at that time was small, and in 1914 he was again writing to the British Museum: *'I have been working on the radium content of meteorites and . . . am applying this to the meteoric analogy of the Earth's internal constitution . . . I have now used up the few meteorites which were in my possession, but I am hoping to get a government grant soon with which to buy others. I feel sure the results will have an important bearing both on the origin of meteorites and the inner structure of the Earth.'* Alas, Holmes did not get his government grant, overtaken as he was by global events and the outbreak of the First World War.

We now know that Chamberlin's Planetesimal Hypothesis was incorrect with regard to the formation of the Earth. Current understanding considers that the whole Solar System formed from a great cloud of gas, most of which collapsed to form the Sun, leaving a small amount of material behind (about one per cent) which condensed to form the planets and other extraneous bodies such as moons, asteroids and comets that subsequently fragmented to create meteors. (Meteors only become meteorites when they hit the Earth's surface.) But regardless of which theory Holmes believed formed the Solar System, in 1912 he recognised that a study of meteorites could hold the key to understanding the early evolution of the Earth because: *'Meteorites allow us to read at our leisure many of the secrets which are otherwise locked up in the Earth's interior'*. His radium study provided us with the first real indication that the Earth may have originated from the same material as other extraterrestrial bodies but, as is often the case, that work lay forgotten when, over forty years later, the genetic link between the Earth and meteorites became an issue of major importance.

As you look up to the stars tiny photons of light that have been travelling through space, perhaps for billion of years before the Earth was even formed, find a resting place in your eyes. Everything that has ever happened in the world, the solar system and the Universe, has occurred in its own time and place just so that that particular photon can reach you at that precise moment in time, millions, and perhaps billions of years after it left its point of origin. The longer you stand there looking up, the further back in time you can see as your eyes adjust to the darkness and more distant stars are revealed. If you stand there long enough you might be lucky and see a shooting star, an extraterrestrial body burning its way through our atmosphere which, if it is big enough, may survive to land here on Earth.

Meteorites hold a special place in the heart of many geologists – the only chance we shall ever have of touching something off-world, something alien and ancient, perhaps from another planet – so over the years many people have analysed meteorites and studied their chemical composition. Fritz Paneth, whom we last encountered working in Berlin on minute amounts of helium, at first had considerable success dating meteorites by the helium method. Unlike terrestrial samples, meteorites did not appear to loose helium, and some very ancient dates were obtained. Eventually however, it was realised that as these bodies cruised the heavens they were bombarded with cosmic radiation that *increased* the amount of helium present, resulting in ages that were anomalously *high* – somewhat ironic after all the problems helium had previously caused with values being too low. Paneth and others also analysed the chemical composition of meteorites and wondered, as Holmes had done, about their genetic relationship to the Earth, but it was not until 1947, just after Holmes had finished his work on the age of the Earth using the Invigtut galena to represent primeval lead, that someone else suggested a better place to look for primeval lead might be in iron meteorites.

It is sad to discover how much the final stage of dating the age of the Earth owes to people involved in the Manhattan Project and development of the atom bomb that killed so many thousands. In the immediate aftermath of dropping the bomb on Hiroshima, while the world waited for Japan's response to the threat of a second one if surrender did not come within forty-eight hours, newspapers carried such headlines as: 'Most terrifying weapon in history: Churchill's warning', and 'Blind girl 'saw' the first flash' (120 miles away). They reported how the project had spent five billion pounds 'on the greatest scientific gamble in history', and proudly claimed that 'British and American scientists have won the race to harness the basic power of the universe in a war weapon with a force which will turn the course of history'. Little did Marie Curie know when she started her investigation of Becquerel's mysterious 'uranium rays' just what it would eventually lead to. But during the Second World War many world-famous scientists and Nobel Laureates were amongst the hundred and twenty-five thousand people who worked on the Manhattan Project. Amongst these were Harrison Brown and his future research student, Claire Patterson, who were both then based at the University of Chicago.

Brown had very wide ranging interests, having been exposed to the challenging environment created by those he worked alongside on the Manhattan Project, but perhaps his greatest talent lay in his flair for extracting the best from others, and teaching them how 'to seek the hidden splendours of science, damning the risks'. After the war Brown stayed on at Chicago and initiated a number of new projects in fields such as nuclear chemistry, chemical oceanography and astrophysics. In addition to all this, and following his involvement with radioactive decay systems on the Manhattan Project, Brown collaborated with American geologists who, following Holmes' work, were dreaming the dream of finding an accurate method of dating common igneous rocks by analysing the impossibly small amounts of lead they contained.

Enthused by their ideas, and recognising Patterson as a skilled mass spectrometrist (and with equipment and levels of funding never available to Holmes) Brown set Patterson to work on determining techniques for the measurement of the tiny quantities of lead found in common igneous rocks. Initially Patterson was severely hampered by problems with lead contamination which, he eventually discovered, came from the atmosphere. But after several years of working with super-clean micro-chemical and mass spectrometric techniques, Patterson and his colleagues were finally able to determine accurate and repeatable lead ages on common igneous rocks. In the early 1950s technology was at last catching up with the dream, and the possibility of using uranium–lead isotopes as a routine tool for dating rocks finally came within their grasp. A further result of this work was Patterson's subsequent campaign to reduce the amount of lead in petrol, for it was Patterson who first drew the world's attention to lead pollution of the atmosphere by emissions from car exhausts, when he realised it was causing the contamination in his lead samples.

Having succeeded in dating common igneous rocks, and as a natural extension to his interests in astrophysics and geochronology, Brown next wondered if the age of the solar system could be determined by finding the age at which meteorites had formed. So in 1952, when Brown and Patterson moved to the California Institute of Technology, Patterson built his first ultra-clean lead laboratory and began the work of determining the composition of lead isotopes in iron meteorites.

The advantage of choosing iron meteorites was that the amount of uranium they contained was negligible, so the primeval lead they held could never have been contaminated by radiogenic lead. The disadvantage was that the amount of lead available for measurement was vanishingly small. Nevertheless, in 1953 Patterson succeeded in determining the lead content of the Canyon Diablo meteorite, a huge extraterrestrial body that had collided with the Earth about fifty thousand years ago, carving out the Meteor Crater in Arizona, which is half a mile wide and 200 metres deep. The Canyon Diablo meteorite contained the lowest lead ratios ever measured. As the enormous significance of this work dawned on Patterson he suggested that these latest meteorite lead values might be a more realistic record of the elusive composition of primeval lead than that found in the Invigtut galena from Greenland.

Fiesel Houtermans, co-founder of the Holmes–Houtermans model for dating the age of the Earth, was quick to pick up on the new lead data, immediately recognising its potential for improving the age of the Earth. Using Patterson's new values from meteorites to represent primeval lead on Earth, instead of the Invigtut galena, and the data from ten Tertiary galenas to represent present day values, he refined his earlier calculation for the age of uranium. He found the new age of the Earth to be 4500 million years, plus or minus 300 million. In one giant leap the Earth had grown older by more than a billion years. But once again Houtermans was just pipped at the post. At a conference held in Wisconsin in September 1953, three months before Houtermans published in December 1953, Patterson presented the results of calculations that were virtually identical to Houtermans. They included two ages for the Earth derived by using his lead values from the Canyon Diablo meteorite, together with modern day values from both a granite and a basalt. These were, respectively, 4510 million and 4560 million years. In age of the Earth terms, they were the same number.

There has always been some debate as to who actually published their results first, Houtermans or Patterson. Back in 1904 when there was some dispute as to who was the first person to date the age of a rock by radioactivity, Rutherford or Strutt, the date of publication of results determined the winner, and in that case it was Rutherford. But in this case, the precise publication date for the conference volume in which Patterson's results were printed is somewhat obscure, the editor believing 'it was probably published in late 1953 or early 1954'. Late 1953 would probably mean before Houtermans, but early 1954 would quite clearly be after him. However, in November 1953 Patterson attended another conference, this time in Toronto, and again reported his age of the Earth results. There a reporter from *Chemical and Engineering News* was in the audience and in the next issue of that journal, circulated on November 23rd, headlines proclaimed, 'Earth's Age: 4.6 Billion Years'. While not strictly the publication of Patterson's results, since Patterson had not written the article himself, it is clear evidence that Patterson had presented his data to an audience of his peers before Houtermans did. But whatever the arguments about the dates of publication, since Patterson developed the techniques for determining these extremely small amounts of lead and the results had been determined by him, whereas Houtermans only used the data, it is fitting that Claire Patterson's name goes down in the history books as the man who finally dated the true age of the Earth. Wild miracle, finally achieved.

Only one question now remained.

What neither Houtermans nor Patterson had shown was whether it was realistic to genetically relate the Earth to meteorites by assuming that primeval lead on Earth could be represented by that found in meteorites. If there was no genetic relationship and the Earth and meteorites had not formed at the same time from the same material, then the primeval lead of meteorites would not be that of the Earth; thus there would be no point in trying to determine the age of Earth from meteorites, and everyone would be back to square one. It was something of a Catch 22 situation: identical ages for the Earth and meteorites indicated a strong genetic link, but the age could only be confirmed if a genetic link was proven.

Patterson spent the next three years trying to prove the relationship and in 1956 he showed the world that the Earth and meteorites did indeed have a common ancestry and that 'The most accurate age of the earth is obtained by demonstrating that the earth's uranium–lead system belongs to the array of meteoritic uranium–lead system'. He had analysed the lead

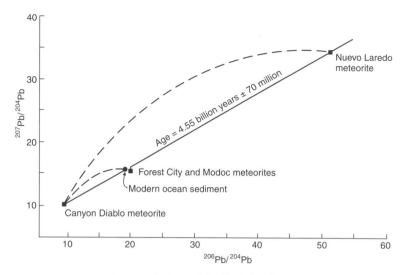

Patterson's Age of the Earth isochron.
Patterson showed that ocean-floor samples fell on the same
uranium–lead time line (isochron) defined by the meteorites
(squares). This meant that both the Earth and meteorites must be
4.55 billion years old, and must have formed from the same
cosmic material. If either had started out with different primeval
lead values they would not now fall on the same line. The dashed
lines show how the lead isotope values have evolved over time
since the 'fixing' of primeval lead in the Universe, as represented
by the Canyon Diablo meteorite.

content of three stony and two iron meteorites and showed how
the values for all five, despite large differences in their long
history, fell on a straight line (isochron) which defined an age
of 4550 million years, give or take 70 million. This proved that
all five meteorites had the same age, but unless values from the
Earth also fell somewhere along that line, a coeval origin would
not be confirmed. But what would be suitable to represent a single sample from the Earth, with its hugely diverse rock types?

Patterson had a brilliant idea. He reasoned that because deep-
sea sediments accumulating on the ocean floor must represent

a wide volume of rock eroded from the continents they had been removed from, they were the most likely to contain average lead values representative of the Earth's crust. Taking samples from the floor of the Pacific Ocean, he analysed the lead ratios and showed that they too fell on the line defined by the five meteorites. This finally proved that all had the same age and the Earth and meteorites were indeed formed at the same time from the same solar material. It must have been a stunning moment. The wildest of all wild miracles.

———————————————— ⚜ ————————————————

So, according to Patterson, the 'time since the earth attained its present mass' was four thousand, five hundred and fifty-five million years, give or take a few million. Within a month of Patterson publishing his results, Holmes gave us his last contribution to the age of the Earth debate, in an article entitled 'How Old is the Earth?' In the ten years since he had last estimated the age of the Earth using Nier's data a great many new analyses of terrestrial leads had become available and, despite his early work on meteorites, Holmes argued, as others did, that 'to use the isotopic composition of lead from iron meteorites as part of the basic data for calculating the age of the earth or its crust, is unsound in principle ... the correct procedure is to use terrestrial materials'. Accordingly he wrote: 'My own attempt to solve the problem from terrestrial evidence alone leads to essentially the same result, which may be expressed as 4,500 ± 100 million years', his larger error encompassing Patterson's results.

Today Claire Patterson is recognised as the man who finally dated the age of the Earth, although a new consensus is developing that the lead isotope clock of the Earth may have been reset by formation of the Earth's core as late as 4500 million years ago, which is significantly younger than the age of most meteorites, now known to be 4560 million years old. If this turns out to be the case, then Patterson's results were more fortuitous than was realised fifty years ago. But at the time Holmes generously and enthusiastically proclaimed the team's achievements: *'this brilliant joint research owed its almost miraculous success to the development of analytical techniques of the utmost precision and delicacy'*. He graciously concludes: *'it is a pleasure to record my indebtedness to many younger friends who are now boldly accepting the challenge and meeting it with all the resources and superb techniques of the atomic age.'* A gentleman to the end, and perhaps comfortable in the knowledge that his name is inextricably linked with the age of the Earth, Holmes did not seem to mind that, after fifty years of dedication to the problem, he did not get there first.

It is evident that while completing his own last words on the subject Holmes had been writing to Patterson who sent him copies of his latest lead results *'in advance of publication'*. Patterson greatly admired Holmes and prized his 1927 'sixpenny' edition of Holmes' book on *The Age of the Earth*. Holmes records that they had *'a little friendly disagreement here and there* [which] *will not, I trust be taken amiss.'* It was probably over whether it was realistic to use the Canyon Diablo meteorite to represent primeval lead, but unfortunately the correspondence that passed between them is now lost. They must have been remarkable documents. In the only letter that remains, Patterson wrote to Holmes enclosing a paper Holmes had asked for. With genuine humilty Patterson took the opportunity to pay his respects to the master:

I wish to reiterate my personal indebtedness to your pioneering work in this field. It was outstandingly inspiring and ingenious.

The Dating Game was over and Arthur Holmes walked off with the very last trophy.

Loose Ends

I don't pretend to understand the universe –
it is a great deal bigger than I am.

Thomas Carlyle

The Earth is very old – present estimates put it at 4.54 billion years, ±45 million. Most of the rocks we see today have been recycled many, many times. They have been down to the bottom of the deepest oceans, buried kilometres below the surface of the Earth, before being uplifted once again to form the very highest peaks of the Himalayas, the Andes, the Rockies or the Alps, where erosion starts them on their weary way again, back down to the sea. The cycle goes on unceasingly. It has done so for billions of years in the past and will continue for billions of years to the future. At the same time the continents have moved around the globe, effortlessly, like so many birds on migration – once buried under glaciers at the poles they soon find themselves passing the equator *en route* to another destination.

Given all this mobility, it is hardly surprising that initially Holmes did not find very ancient rocks on Earth. Most of the evidence has disappeared long ago – but not all. In 1915 he predicted *'It was in zircon that the hope of the future lay, for that mineral was widespread in time and place, and stable and resistant to external forces.'* Indeed, he was right; crystals of the mineral zircon have been found in Western Australia that, at more than four billion years old, are only a few hundred million years younger than the age of the Earth. They are the oldest things so far found on this planet. Today, the oldest rocks in the world are found in the Slave Province of Canada. They are just less than four billion years old, but who knows what might turn up tomorrow?

Moon rocks turned out to be much the same age as Earth rocks, the oldest of these being just over four billion years, providing strong evidence that the Earth and moon were formed at the same time from the same material, although even now the origin of the moon is not entirely clear. In Kelvin's time George Darwin, older son of Charles Darwin and a great supporter of Kelvin's theories of a young Earth (much to his father's annoyance), proposed that the rapid rotation of Kelvin's 'molten globe' had resulted in a huge tidal bulge which eventually spun off from the Earth to form the moon. Back in 1913 Holmes described this theory as *one of the most fascinating romances in the domain of cosmogony'*.

More recently it was thought that the moon was 'captured' by the Earth from a different part of the Solar System, but the very latest idea is that early in the Earth's formation it collided with another body of similar composition, but a third of its size. The huge impact caused the iron core of the two bodies to coalesce while lighter material was projected into space where it later aggregated to form the moon. In particular, this idea explains why the Earth has an unusually dense core, something Holmes speculated upon in 1911 while in Mozambique, and that the moon has no core at all. But whichever way the moon was formed, the fact that the Earth, moon and meteorites all have very similar ages strongly supports the hypothesis that the whole Solar System formed at the same time from the same material.

Following his physical breakdown in 1948, Arthur Holmes returned from recuperating in Ireland in time to attend the International Geological Congress, being held that year in London. In his address from the chair he once again challenged current thinking, this time by turning the accepted geology of Africa upside down: *'The time has come to liberate Pre-Cambrian geology from the tyranny of a telescoped classification'.* Geologists in the audience, such as Robert Shackleton, who was also working on the Precambrian of Africa, remember to this day a flash of revelation as Holmes re-drew the geological map of Southern and Central Africa, based on a relatively small number of radiometric dates. *'No one could be more aware than I am how few and ... how unsatisfactory are most of the age determinations already available. Nevertheless, poor and few though they be, they are unlikely to place a group of rocks in its wrong cycle and they serve to show what far-reaching conclusions can be drawn.'* Holmes made geologists sit up and realise, probably for the first time, how important radiometric dating was in revealing the relationship of one rock to another, thereby illuminating the geological processes that had gone on in the dim and distant past: *'Obviously, all previous correlations not based on absolute dating have been no more than "shots in the dark", and if occasionally they scored a bull's eye it was only by chance.'* Finally, he could not resist a parting shot: *'When maps such as the one here presented are completed for all the major Pre-Cambrian areas ... they will provide a reliable criterion for testing the continental drift hypothesis.'*

Such was the impact on the geological community of Holmes' work on Precambrian geology that it ultimately resulted in major government funding for the establishment of laboratories for geological age determinations in Britain, France, and Belgium, but they were not completed in time to be of value to him. Holmes spent much of the remainder of his working life trying to unravel the mysteries of the Precambrian, whose rocks represent eight-ninths of geological time.

In this connection Alfred Nier continued to date rocks for Holmes, until Nier's interests turned him in another direction and, despite eventually having eight mass spectrometers at his disposal, he was unable to dedicate any of them to analysing lead. Holmes went elsewhere, to Tuzo Wilson's laboratory in Toronto, but a few years later Nier visited Holmes in Edinburgh and the two finally met. In 1960, in the last letter he wrote to Holmes, Nier tells how he had been asked to give an illustrated after dinner speech to geologists:

Included was the delightful photograph made in your garden in July 1954 when Mrs. Nier and I visited you and Mrs. Holmes. Those who had never met you were delighted to see what the 'father' of geological time scales looked like!

With his health progressively deteriorating, Holmes began to have attacks of auricular fibrillation. Although not life-threatening, the very rapid and irregular contractions of the muscle fibres of the heart were exhausting, coming on without warning and putting him out of action for several days. For a while he worked only in the mornings, having to rest every afternoon to decrease the frequency of the attacks. But eventually, in 1955, he submitted his resignation to the Crown and retired from the University in 1956. However, despite his complaints to Reginald Daly about bureaucracy, he did not exactly succumb to administrative over-load, and his passion for research never diminished. In his last years at the University he frequently delegated his administrative duties to other members of staff, or simply did not turn up to meetings, such that a special minute adopted at a Senate meeting at the time of his retirement included the telling words:

To him life is work and work to him means geological research. If his seat at times has been vacant at Senatus, his absence must be weighed against his contributions to science.

In the year of his retirement the Geological Societies of both London and America presented this physicist-turned-geologist with their highest awards for his 'outstanding accomplishment

Arthur Holmes, the 'Father' of geological time scales.

in geology', and in 1964 he was honoured with the greatest accolade a geologist can have – the Vetlesen Prize – for his 'uniquely distinguished achievement in the sciences resulting in a clearer understanding of the Earth, its history, and its relation to the universe'. Characteristically both modest and forthright in his acceptance letter, he expressed his surprise at being selected *for what must surely be the highest distinction in the world for geologists. The surprise was all the greater because I have to confess that I had not even known there was such an award*! The prize had been initiated in 1960 and since it was given only every two years, Holmes was but its third recipient. Its founder had hoped that in time 'this prize will rank in dignity and significance with the Nobel Prizes'. Alas, this does not seem to have come to fruition.

In 1962 the Holmes had moved to a flat at the bottom of Putney Hill in London because the hill on which they lived in Edinburgh *'keeps me a prisoner unless I hire a car'*. No longer capable of travelling, Holmes was unable to attend the prize-giving ceremony for the Vetlesen award being held in America, so the Award Committee brought America to him. At a small ceremony in the Royal Society apartments in London, the gold Vetlesen medal was presented to him along with a sizeable cheque. Arthur Holmes, now a small, frail old man, gave a short speech of thanks:

Looking back it is a slight consolation for the disabilities of growing old to notice that the Earth has grown older much more rapidly than I have – from about six thousand years when I was ten, to four or five billion years by the time I reached sixty. But it is a greater consolation to find that one's work has not gone unappreciated. I have had my share of honours, but I have not deserved these rewards unaided. My wife has been a daily and never-failing source of inspiration and encouragement.

After luncheon he returned home for his afternoon sleep, his beloved wife, '[who] *is probably more worthy than I am to receive so glittering an award'*, by his side.

On his retirement Holmes immediately set out to revise 'Holmes', his *Principles of Physical Geology*, which he considered was long overdue. In 1957 he records that *'I am now getting on with it as quickly as possible'*, but a year later is complaining that *'Every page seems to need a considerable literature search, so much has been done since the war.'* With his failing health, it was indeed a mammoth task, taking him the rest of his life. Almost three times its original length, he finished the second edition just a few months before he died. It was as if he felt he could let go, now his contribution to posterity was finished, his obligation to his students finally discharged.

Arthur Holmes died of bronchial pneumonia at Bolingbroke hospital in London, on the 20th of September, 1965. The small gathering at his cremation a few days later did not even include his son Geoffrey, who at that time was living with his family in Geneva and was told by Doris 'not to bother'. Although only nine years younger than Arthur, Doris lived for another twenty years, until she was eighty-six. In her seventies she bought a car, learnt to drive and wrote yet a third edition of 'Holmes'. She was a remarkable woman. A student of hers, Donald Duff, wrote the fourth edition of 'Holmes', still in use today.

Of Holmes' other great achievements, I have said little, but finally they too are being recognised. Naomi Oreskes, writing on why Americans in particular had such difficulty in accepting continental drift, at last gives Holmes the full credit he deserves for his visionary work on convection currents in the mantle as a mechanism for driving continental plates. She writes:

The pieces of the puzzle [of continental drift], as geologists understand it today, were assembled by the end of the 1920s. Geologists had a phenomenon, they had evidence, and they had a mechanism. What was hailed as breakthrough knowledge in the 1960s . . . was proposed in the

1920s. Holmes' theory was not complete by present standards, to be sure, but neither is existing knowledge complete.

Holmes' theory may not have been complete in the 1920s, but by 1944 in the chapter on continental drift he had so hesitated to include in his *Principles of Physical Geology*, his ideas come very close to being *'the all-embracing theory which would satisfactorily correlate all the varied phenomena for which the earth's internal behaviour is responsible'*. But while recognising continental drift as the all-encompassing theory it would become, he nevertheless found it necessary to caution his students about such speculation: *'It must be clearly realised, however, that purely speculative ideas of this kind, specially invented to match the requirements, can have no scientific value until they acquire support from independent evidence.'* Little did he realise that he himself had already pioneered the means by which that 'independent evidence' would be found.

During the Second World War, sensitive instruments were developed to detect submarines by their magnetic fields, but as they searched backwards and forwards across the oceans an extraordinary picture of the sea floor began to emerge. Linear bands of positive and negative magnetic anomalies, some stretching hundreds of miles along the ocean floor, were seen to show an almost perfect symmetry either side of mid-ocean ridges – large 'cracks' or fissures that extended down the middle of all the great oceans. This peculiar magnetic pattern puzzled scientists for years, until in 1963 the startling proposal was made that it represented bands of rock on the sea floor that were magnetized during normal and *reversed* periods of the Earth's magnetic field.

For some time it had been recognised that when rocks such as basalt cooled from their molten state, the magnetic particles within them became 'fossilised', pointing in the direction of the Earth's magnetic field. Thus it was proposed that the stripes on the ocean floors represented magnetised basaltic flows that had erupted from the fissures along the mid-ocean ridges, acquiring the prevailing direction of magnetisation as they cooled. But, and this was the crucial point, if every now and then the polarity of the Earth's magnetic field flipped over, then the material being extruded at that time would acquire a *reversed* magnetisation. Over millions of years, the sea-floor basalts would record the orientation of the Earth's magnetic field like a bar code. Furthermore, it was argued, these bands could be used as evidence in favour of the new theory of 'sea-floor spreading' that had been put forward a few years earlier.

Supporters of sea-floor spreading proposed that as molten magma welled up from the depths along the mid-ocean fissures, it pushed the previous flow apart so that half would move to either side of the ridge, slightly widening the ocean each time and eventually creating new sea floor. After this had occurred many, many times the original flow could finally end up thousands of miles away from the ridge, and the two continents, originally part of the same landmass, would be thousands of miles apart.

Although this was an extraordinary concept, once it was proposed evidence for reversals of the Earth's magnetic field throughout geological time was rapidly found in thick onshore lava flows, and the 'flipping' of the earth's magnetic field quickly became an accepted fact. But while magnetic reversals certainly seemed to explain the anomalous stripes seen either side of ocean ridges, thus supporting the theory of sea-floor spreading, without 'evidence' the theory remained, like continental drift, locked in the realms of speculation.

The evidence was found when they learnt how to 'read' the bar code – and this was done by *dating* the stripes. The isotopic system used for this dating was the decay of potassium to argon. In 1932 Holmes had hoped that one day the decay of potassium to calcium would revolutionise his work, but it was not until 1948 that Alfred Nier demonstrated the dual decay of potassium to both calcium *and* argon, and it was the latter which would be used for dating purposes.

By dating samples of rocks from onshore lava flows where magnetic reversals had been identified, a *time scale* of magnetic reversals was gradually built up until eventually it became possible to 'read' the ages of the reversals on the sea floor. Immediately it became clear that the youngest rocks were nearest to the ridge, while the oldest were furthest away and adjacent to the continents. Either side of the ridge, stripes of exactly the same age could be matched up with one another. So the oceans really were opening, and the continents really were drifting apart. Wild miracle confirmed! Continental drift at long last became a reality.

Within a couple of years, and certainly by the end of the 1960s, a revolution had occurred in the earth sciences, and there were only a handful of geologists left who still did not accept the 'new' ideas about sea-floor spreading and continental drift. The key to that revolution was development of a geological time scale.

Today geology has its dates, just like history does. Time has become the framework onto which we hang all geological events and, as in our daily lives, it has become indispensable. We have learnt how to tell geological time from isotopic clocks, and we have developed a time scale for the evolution of life, and all that went before it. Using the clocks and the time scales we have discovered the true age of 'Mother Earth', revealed many of her internal mysteries and developed a unifying theory that explains all geological processes – just as Arthur Holmes' vision of 1913 said we would:

With the acceptance of a reliable time-scale, geology will have gained an invaluable key to further discovery. In every branch of the science its mission will be to unify and correlate, and with its help a fresh light will be thrown on the more fascinating problems of the Earth and its Past.

Given enough time, everything possible happens.

Thanks and Acknowledgements

Very few things happen at the right time, and the
rest do not happen at all: the conscientious historian
will correct these defects.

Mark Twain

If gratitude could be measured on the geological time scale, then
what I owe Hugh Torrens, historian of technology and Professor
of Geology at Keele University, would stretch beyond the age of
the dinosaurs and survive the Permian extinction, only to disappear down the black hole of the Archaean still feeling inade-
quate. Without the benefit of his invaluable advice this novice
historian would have taken aeons to find all the relevant materials, and this book would probably still be sitting on a word
processor.

Finding out about another person's life is like writing a detec-
tive story – except that you are in it. Arthur Holmes left few clues
about his private life and the 'garden shed mystery' was never fully
resolved. A shed at the bottom of Doris' garden in Hove was
believed to contain all Holmes' papers, but quite what happened
to its contents when she died is not clear. I found some, but certainly not all. Although Geoffrey Holmes, Arthur's son, sadly died
before I had a chance to meet him, I was delighted when I finally
tracked down Geoffrey's wife Karla, and their four children. They
very kindly provided me with access to Arthur's letters from
Mozambique, as well as several of the early photos of Arthur

Holmes, his family and travels. Jill Reynolds, niece of Doris Reynolds, gave me some wonderful photos of Doris and a fascinating insight into what she was like as a person.

When I first rang up Edinburgh University trying to locate Holmes' diaries which I had seen a vague reference to in Doris' obituary of her husband, I was advised to contact Professor Gordon Craig. Following my enquiries his first email said:

Subj: Arthur Holmes

Good to hear from you. If you are serious about working on Holmes you had better come to Scotland. I was his first Edinburgh appointment. CDW Waterston was one of Holmes' students, DB McIntyre was another. Fred Stewart, who succeeded Holmes in the Regius Chair lives in Argyll. Miss CGM Berry (now living in Kirkcaldy) was Holmes' secretary and although now in her 80s has still a first-class memory for people and events. All of them I am sure will be happy to talk to you.

Good luck, Gordon Craig

I took Gordon's advice and immediately arranged a visit to Scotland to meet everyone he suggested. True to his word they were all happy to see me, and I cannot thank him and them enough for the memories and insights they shared with me, particularly the remarkable Miss Berry, whose memory for detail was truly staggering. Gordon also led me to Leslie Black and Jean Duff, wives of two other of Holmes' students who were good enough to give me access to more Holmes-related material, as well as to Grace Dunlop, who had been on Holmes' staff in Edinburgh. She told me the story of Holmes' near-death experience in Mozambique and that he was 'a most considerate and polite gentleman . . . keen to encourage women to go forward in the profession'. Gordon also located the wonderful diaries for me. These will form part of an Arthur Holmes Collection that is being established at the Geological Society in London. It is quite scandalous that one has not already been formed.

The Durham period was harder to follow up, but the redoubtable Sir Kingsley Dunham had a store of memories he willingly shared with me. He allowed me to quote him verbatim because I could not say it better myself. Mary Wilson and her sister Dorothy Edmondson, kindly wrote to me when they heard I was looking for people who knew Arthur Holmes. Being the daughters of Maggie Holmes' best friend, they were able to tell me what little we now know about Maggie. They also filled a crucial gap in the story – what happened to the Holmes when they returned from Burma? Arthur's subsequent job applications always implied that he had returned from Burma to take up the Durham post, but that did not seem to square with the diary, which indicated an eighteen month gap. Now I know better!

So far I have not been able to trace any of Bob Lawson's relatives, so if any of you are out there, please do get in touch.

Many archivists and librarians, particularly Beth Rainey at Durham University Library and Ian Stewart and Kate Barthel in the Durham University Office, Vanna Skelley at Burmah Oil, Sophie Badham at Royal Holloway, Anne Barrett at Imperial College and Wendy Cawthorne at the Geological Society were immensely helpful in my search for Holmes-related material, but I shall particularly remember John Thackray, archivist at the Natural History Museum, who wrote telling me they held about sixty letters from Arthur Holmes. I could not get down to London fast enough. Those letters provided crucial information on Holmes' early work on meteorites. Tragically John died before being able to see my thanks in print.

It was while writing this book that I discovered the wonders of email. Writing is a lonely business, so it helps a great deal to converse with people all over the world without having to leave your desk. I made many American friends that way, particularly Naomi Oreskes, Hank Frankel and Alan Allwordt, who all share my admiration for Holmes, and who greatly helped with sourcing material on that side of the Mid Atlantic Ridge. There were many others, from India to Australia, who wrote to me with information about Holmes. Everyone praised the man or his work, and I was grateful for it all, helping me as it did to build up a picture of this quiet and unassuming individual.

Stephen Moorbath, Norman Snelling, Tony and Brenda Hurford, members of my family and several anonymous reviewers were recruited to read the manuscript, putting the science right where it had gone wrong and making many other helpful improvements. I refused, however, to change the title, which arose from a session in the pub with two Antipodean colleagues, Geoff Laslett and Rex Galbraith, so it is all their fault if you found this book on the 'Romance' shelves of your bookshop and were surprised by its contents.

Finally and most importantly I must thank Robert Shackleton who, if unwittingly, first introduced me to Arthur Holmes. Little did he, or I, anticipate what it would eventually lead to.

To you all I can but say 'thanks a million'.

Permission to quote from various letters and sources has been generously granted by the following individuals or institutions.

Quote on page

Courtesy of Leslie Black:
    Letter to George Black from Doris Holmes,
      24th October, 1960.   213

Courtesy of the Burmah Castrol Archives:
    Letter to Finlay Fleming & Co. from R.I. Watson,
      MD Burmah Oil, 15th April, 1920.   123

Courtesy of Mrs. Lorna Patterson and the California Institute of Technology:
    Letter to Arthur Holmes from Clair Patterson,
      12th July, 1963.   228

Courtesy of the Chicago University Library:
    Letter to T.C. Chamberlin from Arthur Holmes,
      4th June, 1912.   217

Courtesy of the University of Durham:
    Joint Board Minutes for 1931–1935, 19th January, 1933.   164
    Minute 237 (h) of the Council Minutes for 30th January,
      1940.   177a

Courtesy of the Edinburgh University Library:
    Letter to W.W. Watts from Thomas Holland,
    15th February, 1943.   186a
    Letter to The Vice Chancellor, Edinburgh University,
    from W.W. Watts, 17th February, 1943.   186b
Courtesy of Henry Frankel:
    Letter to Henry Frankel from Doris Holmes,
    4th February, 1979.   157
Courtesy of the Harvard University Archives:
    Letter from Reginald Daly to Arthur Holmes,
    27th March, 1943.   181
    Letter to Reginald Daly from Arthur Holmes,
    3rd March, 1951.   214a
    Letter to Reginald Daly from Arthur Holmes,
    12th May, 1951.   214b
    Letter to Arthur Holmes from Reginald Daly,
    22nd March, 1951.   214c
Courtesy of John Hepworth:
    Letter to C. Lewis from J. Hepworth, 6th October, 1998.   183
By permission of the Archives of Imperial College of Science,
Technology and Medicine, London:
    Imperial College, History of the Department of Geology,
    by D. Williams, October, 1963.   32
    Departmental Histories: Physics Department, 1962.   57
    Phoenix article by J. Forgan-Potts: The Social Aspect
    of the War, May, 1915.   108a
    Phoenix article by G.S.M. Taylor: In the Trenches,
    May, 1915   108b
Courtesy of University of Minnesota Archives:
    Letter from Arthur Holmes to Alfred Nier, 21st May, 1945. 199
    Letter from Arthur Holmes to Alfred Nier,
    16th February, 1946.   203
    Letter from Alfred Nier to Arthur Holmes, 3rd May, 1960. 232a
By permission of the Trustees of The Natural History Museum:
    Rules and Regulations for post of Assistantship in the
    Department of Minerals, Natural History Museum,
    25th May, 1910.   51

Letter from Arthur Holmes to Dr Prior,
4th February, 1914.   218a
Letter from Arthur Holmes to Dr Prior, 24th August, 1919.   118
Letter from Arthur Holmes to Dr Prior, 4th June, 1924.   141
Letter from Major R. Archer Houblon,
16th December, 1929.   218b
Letter from Arthur Holmes to Dr Claringbull,
10th September, 1957.   235a
Letter from Arthur Holmes to Dr Claringbull,
24th July, 1958.   235b
Courtesy of the Royal Holloway University of London Archives:
Letter to Arthur Holmes from Board of Education,
14th September, 1907.   28
Letter to Dr M.P. Appleby from Arthur Holmes,
24th April, 1940.   177b
Letter to the Governing Board, Edinburgh University, from
Reginald Daly, 29th December, 1942.   186c
Letter to Arthur Holmes from Sir Thomas Holland,
26th November, 1942.   186d
Letter to Arthur Holmes from the Scottish Home Office,
30th March, 1943.   186e
Letter to Arthur Holmes from C. H. Stewart,
19th October, 1956.   232b
Letter to Arthur Holmes from Grayson Kirk,   234a
15th January, 1964.
Letter to Grayson Kirk from Arthur Holmes,
15th February, 1964.   234b
Arthur Holmes response to Vetlesen award,   234c
14th April, 1964   and d
Courtesy of Yale University Library:
Letter from William Bowie to Charles Schuchert,
11th October, 1928.   158

Permission to reproduce photographs, and other illustrative material has been kindly given by the following individuals or institutions:

Courtesy of Durham University Geology Department:
    Photo of Arthur Holmes, Professor of Geology at Durham University.
Courtesy of Gateshead Library:
    Photo of Gateshead High School
    Photo of Gateshead 1900
    Photo of Ironmonger's Shop
With permission of the Geological Society London:
    Photo of Arthur Holmes, the 'Father' of geological time scales
Courtesy of Karla Holmes and family:
    Photo of Bob Lawson
    Photos of Mozambique
    Extracts from letters to Holmes' parents and Bob Lawson
    Photo of Lim Chin Tsong's Palace
    Photo of Yenangyaung Oilfield
    Photo of Norman Holmes
    Photo of Arthur, Maggie and Geoffrey Holmes, Durham c. 1930
    Photo of Doris and Arthur on the Giant's Causeway, Northern Ireland
Copyright Shropshire County Museum Service, original illustration is in the collection at the Ludlow Museum. Photographer Gareth Thomas, FRPS:
    The Dhustone Section
Courtesy of Jill and Ian Reynolds:
    Photo of young Arthur Holmes
    Photo of Doris Reynolds at about the time she met Arthur Holmes
By permission of the President and Council of the Royal Society:
    Photo of Kelvin as an old man
Courtesy of the Royal Geographical Society:
    Map of Memba Minerals Expedition in Mozambique

Selected Bibliography

No doubt true historians will berate me for not having provided them with notes on every comment and quotation, but I personally find them so intrusive that I decided to simply include a selection of further reading for those interested in learning more. I have provided a 'Top Ten' for the more general reader, followed by a more detailed selection for those wishing to follow the subject in some depth. Both lists are ordered chronologically, more or less in the order in which they were used in the text. An (almost) complete bibliography for Arthur Holmes can be found in Dunham's obituary.

Top Ten

1. Holmes, Arthur, 1913. *The Age of the Earth*, Harper & Brothers
2. Holmes, Arthur, 1944. *Principles of Physical Geology*, first edition, Thomas Nelson and Sons Ltd
3. Harper, C.T. (ed.), 1973. *Geochronology: Radiometric Dating of Rocks and Minerals. Benchmark Papers in Geology*, Dowden, Hutchinson & Ross
4. Burchfield, Joe, D., 1975. *Lord Kelvin and the Age of the Earth*, Science History Publications

5. Glen, William, 1982. *The Road to Jaramillo*, Stanford University Press
6. Dalrymple, G. Brent, 1991. *The Age of the Earth*, Stanford University Press
7. Allegre, Claude, 1992. *From Stone to Star: A View of Modern Geology* (translated by D. Kermes van Dam), Harvard University Press
8. Brush, Stephen, 1996. *A History of Modern Planetary Physics*, in 3 volumes, Cambridge University Press
9. Oldroyd, David, 1996. *Thinking about the Earth: A History of Ideas in Geology*, Athlone Press
10. Oreskes, Naomi, 1999. *The Rejection of Continental Drift*, Oxford University Press

Further Reading

Thomson, W. (Lord Kelvin), 1871. On geological time. *Transactions of the Geological Society of Glasgow*, Vol. 3, part 1, pp. 321–329

Haughton, Samuel, 1878. A geological proof that the changes in climate in past times were not due to changes in position of the Pole, with an attempt to assign a minor limit to the duration of Geological Time. *Nature*, Vol. 18, pp. 266–268

Darwin, Francis (ed.) 1887. *Life and Letters of Charles Darwin*, 2nd edition, John Murray

Walcott, C.D., 1893. Geological time as indicated by the sedimentary rocks of North America. *Journal of Geology*, Vol. 1, pp. 639–676

Sollas, W.J., 1895. The age of the Earth. *Nature*, Vol. 51, pp. 533–534

Poulton, E.B., 1896. President's Address, Section D: A naturalist's contribution to the discussion upon the age of the Earth. *Report of the British Association for the Advancement of Science* (66th Meeting, Liverpool), pp. 808–828

Thomson, W. (Lord Kelvin), 1899. The age of the Earth as an abode fitted for life. *The London, Edinburgh and Dublin Philosophical Magazine and Journal of Science*, Vol XLVII, 5, pp. 66–90

Joly, J., 1900. On the geological age of the Earth. *Report of the British Association for the Advancement of Science*, pp. 369–375

Sollas, W.J., 1900. President's Address, Section C: Evolutional

Geology. Report of the British Association for the Advancement of Science, pp. 711–730

Various, 1906. Radium correspondence. *The Times*, August–September

Strutt, R.J., 1906. On the distribution of radium in the Earth's crust, and on the Earth's internal heat. *Proceedings of the Royal Society*, Ser. A, Vol. 77, pp. 472–485

Joly, J., 1909. *Radioactivity and Geology*. Archibald, Constable and Co.

Holmes, Arthur, 1911. The association of lead with uranium in rock-minerals, and its application to the measurement of geological time. *Proceedings of the Royal Society*, Ser. A, Vol. 85, pp. 248–256

Joly, J., 1911. The age of the Earth. *Philosophical Magazine*, Vol. XXII, pp. 357–380

Holmes, Arthur & Wray, D.A., 1913. Mozambique: a geographical study. *Geographical Journal*, Vol. 42, pp. 143–152

Soddy, Frederick, 1913. Intra-atomic charge. *Nature*, Vol. 92, pp. 399–400

Holmes, Arthur, 1914. The terrestrial distribution of radium. *Science Progress*, Vol. 9, pp. 12–36

Joly, J., 1914. The birth time of the world. *Science Progress*, Vol. 33, p. 37

Holmes, Arthur, Evans, J.W., Young, A.P. & Shelton, H., 1915. Abstract and discussion on: Shelton, H. S. The Radioactive Methods of Determining Geological Time. *Abstracts and Proceedings of the Geological Society*, Vol. 971, pp. 63–66

Holmes, Arthur, 1915. Contribution to the discussion on radioactive evidence of the age of the Earth. *Report of the British Association for the Advancement of Science* (85th Meeting, Manchester), Section C – Geology, pp. 432–434

Holmes, Arthur, 1920. The measurement of geological time. *Discovery*, Vol. 1, pp. 108–114

Strutt *et al.*, 1921. Joint discussion on The Age of Earth. *British Association for the Advancement of Science*, pp. 413–415

Chamberlin, T.C., Clarke, J.M., Brown, E.W. & Duane, W., 1922. From the geological view-point; from the paleontological view-point; from the point of view of astronomy; the radioactive point of view. *Proceedings of the American Philosophical Society*, Vol. LXI, No. 4, pp. 247–88

Holmes, Arthur, 1926. Radium uncovers new clues to earth's age, *New*

York Times, 6 June, Sect. IX, p. 4f, In: Sullivan, W. (ed.) *Science in the Twentieth Century*, Arno Press (1976), pp. 175–177

Holmes, Arthur, 1927. *The Age of the Earth*, 2nd edition. Harper & Brothers

Holmes, Arthur & Lawson, Robert W., 1927. Factors involved in the calculation of the ages of radioactive minerals. *American Journal of Science*, Series 5, Vol. 13, pp. 327–344

Holmes, Arthur, 1928. Continental drift. [Review of the 1926 *Symposium Proceedings on Continental Drift*]. *Nature*, Vol. 122, pp. 431–433

Holmes, Arthur & Dubey, V.S., 1929. Estimates of the ages of the Whin Sill and the Cleveland Dyke by the helium method. *Nature*, Vol. 123, pp. 794–795

Holmes, Arthur, 1931. Radioactivity and earth movements. *Transactions of the Geological Society of Glasgow for 1928–29*, Vol. 18 (3), pp. 559–606

Holmes, Arthur, 1931. Radioactivity and geological time. *Bulletin of the National Research Council*, Washington, Vol. 80, pp. 124–459, In: *Physics of the Earth, Part 4: The Age of the Earth*, by the subsidiary committee on the Age of the Earth

Holmes, Arthur, 1932. The origin of igneous rocks. *Geological Magazine*, Vol. 69, pp. 543–558

Holmes, Arthur, 1937. *The Age of the Earth*, 3rd edition. Thomas Nelson & Sons Ltd

Nier, Alfred O., 1938. Variations in the relative abundances of the isotopes of common lead from various sources, *Journal of the American Chemical Society*, Vol. 60, pp. 1571–1576

Eve, A.S., 1939. *Rutherford*, Cambridge University Press

Nier, Alfred O., 1939. The isotopic constitution of radiogenic lead and the measurement of geological time, II. *Physical Review*, Ser. 2, Vol. 55, pp. 153–163

Nier, Alfred O., Thompson, Robert W. and Murphy, Byron F., 1941. The isotopic constitution of lead and the measurement of geological time, III. *Physical Review*, Ser. 2, Vol. 60, pp. 112–116

Holmes, Arthur, 1944. The age of the Earth. [Inaugural lecture, 1943.], *University of Edinburgh Journal*, Autumn 1944, pp. 12–20

Holmes, Arthur, 1946. An estimate of the age of the Earth. *Nature*, Vol. 157, pp. 680–684

Kuenen, Ph. H., 1947. An estimate of the Age of the Earth: a reply to Holmes. *Geological Magazine*, Vol. 84, p. 57

Holmes, Arthur, 1947. An estimate of the age of the Earth: a reply to Kuenen. *Geological Magazine*, Vol. 84, pp. 123–126

Holmes, Arthur, 1947. The age of the earth. *Endeavour*, Vol. 6 (23), pp. 99–108

Holmes, Arthur, 1947. A revised estimate of the age of the earth. *Nature*, Vol. 159, pp. 127–128

Holmes, Arthur, 1947. The construction of a geological time-scale. *Transactions of the Geological Society of Glasgow*, Vol. 21, Part 1, pp. 117–152

Egerton, A.C., 1948. Lord Rayleigh 1875–1947. *Obituary Notices of Fellows of the Royal Society for 1947*, pp. 503–538

Holmes, Arthur, 1951. The sequence of Pre-Cambrian orogenic belts in South and Central Africa. *International Geological Congress (14), Report of the Eighteenth Session, Great Britain, 1948.* pp. 254–269

Houtermans, F. G., 1953. Determination of the age of the earth from the isotopic composition of meteoritic lead. *Nuovo Cimento, Ser. 9*, Vol. 10, no. 12, pp. 1623–1633

Patterson, Claire, 1953. The isotopic composition of meteoric, basaltic and oceanic leads, and the age of the earth. In: *Proceedings of the Conference on Nuclear Processes in Geologic Settings*, Williams Bay, Wisconsin, September 21–23, 1953, pp. 36–40

Campbell Smith, W., 1956. Presentation of the Wollaston Medal to Professor Arthur Holmes, *Proceedings of the Geological Society of London*, Session 1955–1956, Nos. 1530–1541, pp. 95–97

Holmes, Arthur, 1956. How old is the earth? *Transactions of the Edinburgh Geological Society*, Vol. 16, Part 3, pp. 313–333

Patterson, Claire, 1956. Age of meteorites and the Earth. *Geochimica et Cosmochimica Acta*, Vol. 10, pp. 230–237

Hedberg, Hollis D., 1957. Presentation of Penrose Medal to Arthur Holmes. *Proceedings Volume of the Geological Society of America for 1956*, pp. 69–74

Holmes, Arthur, 1960. A revised geological time-scale. *Transactions of the Edinburgh Geological Society*, Vol. 17, Part 3, pp. 183–216

Chadwick, Sir James (ed.), 1962. *The Collected Papers of Lord Rutherford of Nelson*. George Allen & Unwin Ltd

Holmes, Arthur, 1965. *Principles of Physical Geology* 2nd Edition. Thomas Nelson & Sons Ltd

Cahen, L., 1966. Memorial to Arthur Holmes (1890–1965). *Geological Society of America Bulletin*, Vol. 77, Part 7, pp. P127–P135

Dunham, K. C., 1966. Arthur Holmes. *Biographical Memoirs of Fellows of the Royal Society*, Vol. 12, pp. 291–310

King, B. C., 1967. Arthur Holmes. *Proceedings of the Geological Society of London*, Nos. 1629–1636, pp. 196–201

Reynolds, Doris L., 1968. Memorial of Arthur Holmes, January 14, 1890–September 20, 1965. *American Mineralogist*, Vol. 53; Parts 3–4, pp. 560–566

Badash, Lawrence, 1968. Rutherford, Boltwood, and the age of the Earth; the origin of radioactive dating techniques. *Proceedings of the American Philosophical Society*, Vol. 112, Part 3, pp. 157–169

Badash, Lawrence (ed.), 1969. *Rutherford and Boltwood: Letters on Radioactivity*. Yale University Press

White, George (ed.), 1970. *James Hutton's: System of the Earth, 1785; Theory of the Earth, 1788; Observations on Granite, 1794*. Hafner Publishing Company

Burchfield, Joe D., 1974. Darwin and the dilemma of geological time. *Isis*, Vol. 65, Part 228, pp. 301–321

Frankel, Henry, 1978. Arthur Holmes and continental drift. *The British Journal for the History of Science*, Vol. 11, Part 2, No. 38, pp. 130–150.

Dean, Dennis R., 1981. Age of the earth controversy. *Annals of Science*, Vol. 38, pp. 435–456

Nier, A.O., 1982. Some reminiscences of isotopes, geochronology and mass spectrometry. *Annual Review of Earth and Planetary Science*, Vol. 9, pp. 1–17

Brice, W.R., 1982. Bishop Ussher, John Lightfoot and the Age of Creation. *Journal of Geological Education*, Vol. 30, pp. 18–24

Corley, T.A.B., 1983. *A History of the Burmah Oil Company 1886–1924*. Heinemann

Kauffman, G.B., 1986. *Frederick Soddy (1877–1956)*. D. Reidel Publishing Company

Smith, Kirk R., Fesharaki Fereidun, & Holdren, John P. (eds.), 1986.

Earth and the Human Future: Essays in Honour of Harrison Brown. Westview Press

Allwardt, Alan, 1988. Working at cross purposes: Holmes and Vening Meinesz on convection. *Eos*, Vol. 69, No. 41, pp. 899–906

Fröman, Nanny, 1996. Marie and Pierre Curie and the Discovery of Polonium and Radium, Lecture by Nanny Fröman at the Royal Acadamy of Sciences in Stockholm, Sweden, on Feb. 28, 1996, published on the Internet

Gribbin, John, 1996. *Companion to the Cosmos.* Weidenfeld and Nicholson

Bragg, Melvyn and Gardiner, Ruth, 1998. *On Giants' Shoulders.* Hodder and Stoughton

Courtillot, Vincent, 1999. *Evolutionary Catastrophes: The Science of Mass Extinction,* Cambridge University Press